*Building Contracts for
Design and Construction*

And other things of this sort should be known to architects, so that, before they begin upon buildings, they may be careful not to leave disputed points for the householders to settle after the works are finished, and so that in drawing up contracts the interests of both employer and contractor may be wisely safe-guarded.

VITRUVIUS (Book I)

Building Contracts for Design and Construction

Harold D. Hauf

FELLOW, AMERICAN INSTITUTE OF ARCHITECTS
FELLOW, AMERICAN SOCIETY OF CIVIL ENGINEERS

Professor, Department of Architecture
University of Southern California

John Wiley & Sons, Inc.
New York · London · Sydney · Toronto

Library of Congress Catalog Card Number: 68-9246
SBN 471 36002 3
Printed in the United States of America

Preface

This book deals with the contractual relations between the owner and the architect-engineer firm that designs a project and those between the owner and the construction organization that builds it. In view of the expanding concept of professional liability engendered by court interpretations in recent years, considerable emphasis is placed on the role of the architect-engineer in the selection and specification of materials and on his position during the construction phase of a project. An effort has been made to identify the principal contingencies that may arise in these areas and to suggest provisions for handling them that will be equitable to all parties concerned.

Since it is hoped that the book may be found useful by owners and others who initiate or manage building programs as well as by architects, engineers, contractors, and students, the discussion of contracts for design services precedes that of construction contracts. This order seems appropriate in any event because the effectiveness of the construction operation depends so heavily on the quality of the services performed under the design contract.

As a background for the discussion of contract provisions and their administration, an overview of the customary relationships that exist among the various segments of the building industry is presented first. The subsequent analysis of contract documents derives from their character as operational instruments during performance of the work; for example, the several provisions, known as the "general conditions" of the construction contract, have been identified over time as those items that must be covered by prior agreement if disputes are to be avoided.

Various standard or prototype contract documents have been promulgated from time to time by organizations that make up the design and construction segments of the building industry and by government construction agencies. Several of these are cited in the text and reproduced for illustrative purposes.

Although care must be exercised when applying these documents to specific project conditions, they generally represent the best current practice. Developed by the collaborative efforts of architects, engineers, owners, contractors, and legal and insurance counsel, these documents represent a vast amount of collective building experience. Modification of their provisions as printed should be undertaken only with legal advice.

Although detailed discussion of the writing of technical building specifications is outside the purpose of this book, the ever-growing importance of the specification as a contract document warrants inclusion of a chapter treating specifications from this point of view. The traditional relationship between the nontechnical provisions of the specification and the general conditions of the construction contract will undoubtedly become increasingly closer as new techniques of practice emerge in the industry.

Much of the analytical material in the book has been derived from direct experience in the administration of a considerable volume of design and construction work on both private and government projects. Some of it was used originally in a seminar on building industry relationships I gave at Rensselaer Polytechnic Institute and subsequently in a similar course in the Department of Architecture at the University of Southern California. Questions pertaining to the subject matter of individual chapters follow the appendix and are intended to provide food for thought by persons using the book independently as well as to facilitate its use as a textbook.

I wish to express my appreciation to my professional associates in both private practice and government construction agencies who have contributed to this book. Special acknowledgement is made to Harold P. King, CEC, for valuable suggestions concerning relationships between consulting engineers and architects and to Addison Mueller, Professor of Law at the University of California, Los Angeles, who read portions of the manuscript and who has long acted as my legal conscience in these matters through personal association and his own writings.

H. D. HAUF

Los Angeles, California
June 1968

Contents

Services Not Included • Construction Cost of Consultant's Part of the Project • Consultant's Cost Estimate Responsibility • Payments and Reimbursable Expenses • General Terms and Conditions of Consulting Contracts • Ownership of Plans and Specifications • Termination of Consulting Contract • Arbitration of Disputes • Insurance and Mutual Assistance • Non-Engineer Consulting Agreements • Engineer as Prime Professional • Examples of Contract Forms.

Identification of the Contract Documents • Conditions of the Contract as a Separate Document • Articles of the General Conditions • Definition and Status of the Contract Documents • Administrative Functions of the Architect • Rights and Duties of the Owner • Rights and Duties of the Contractor • Subcontracting Procedures • Performance and Payment Bonds • Royalties and Patents • Arbitration of Claims and Disputes • Disputes Clauses in Government Contracts • Contract Time, Delays, and Extensions • Payments to the Contractor • Safety Precautions and Insurance • Changes in the Work • Uncovering and Correction of Work • Termination of the Contract • Miscellaneous Provisions • Separate Contracts • The Supplementary Conditions • Examples of Standard Forms

Nature of the Specifications • Organization: The Uniform System • General Requirements • Types of Technical Specifications • Tests for Specification Compliance • Technological Responsibility • Use of Reference Standards • Closed Specifications • Open Specifications • Combination Type Specifications • Language of the Specification • Use of Scope Statements

Surety Bonds • Performance Bond • Payment Bond • The Dual-Bond System • Surety Bonds and the Contract • Choice of Surety • Bid Bonds • Miscellaneous Bonds and Warranties • Construction Insurance • Contractor's Liability Insurance • Project and Property Insurance • Owner's Liability Insurance • Effecting Insurance Coverage • Examples of Documents

Nature of the Bidding Process • Bidding Documents • Invitation to Bid • Advertisement for Bids • Instructions to Bidders • Proposal Form • The Bid Opening • Canvassing the Bids • Award of Contract • Prequalification of Contractors • Bid Depositories • Separate Bidding • Factors Influencing Level of Bids • Example of Prequalification Form

Nature of the Agreement · Identification of Parties and Project · Identification
of Contract Documents · Contract Time · Liquidated Damages · Contract Sum ·
Progress Payments · Final Payment · Execution of Agreement · Separate
Contracts Agreements · Short Form Contracts · Subcontracts · Examples of
Standard Forms

Underlying Principles · The CPFF Agreement · Identification of Parties and
Project · Identification of Contract Documents · Contract Time · Reimbursable
Costs · Non-Reimbursable Costs · Contractor's Fee · Changes in the Work ·
Subcontracts · Payments to the Contractor · Accounting and Cost Control ·
Miscellaneous Provisions · Management Contracts · Examples of Cost-Plus-Fee
Contracts

A. The Standards of Professional Practice, American Institute of Architects, AIA
   Document J 330.
B. Code of Ethical Conduct for Members of the Associated General Contractors
   of America.
C. Construction Industry Arbitration Rules, American Arbitration Association.
D. Policy Statement on Building Product Development and Uses, American
   Institute of Architects.

Chapter 1

Building Industry Relationships

Components of the Industry

Because of the great diversity of elements and interests that make up the building industry and because of its general looseness of organization as compared, for example, with the automotive industry which is dominated by a few large corporations, there are differing views concerning its organizational structure. Fortunately, agreement is not a requisite to an analysis of the workings of the industry; nevertheless some grouping of the more or less distinct types of businesses involved in construction is essential to understanding their relationships. For the purposes of this discussion the following classification of industry elements will be used:

Design professions
Building constructors
Building product manufacturers
Building finance agencies
Real estate organizations

The first three may be considered the technical segments of the industry and the last two allied or related elements.

Architects, together with civil, structural, mechanical and electrical engineers, city and site planners, landscape architects, and many other special consultants, comprise the design professions. General contractors, mechanical and electrical contractors, and numerous special-trade contractors constitute the principal members of the group who carry out the actual construction of buildings. Producers of building materials and components include an extremely wide range of manufacturing enterprises—from the local brickyard, practically all of

1

whose product is used in local buildings, to the large industrial corporation whose products are marketed nationally and for whom the building market may represent only a portion of its total manufacturing activity.

Financial organizations including mortgage agencies, savings and loan associations, and commercial banks, are less closely related to the physical production of buildings than are the three technical segments of the industry. Nevertheless, the measures that lenders and mortgage-insuring agencies take to assure the quality of the properties they finance often exert a major influence on the planning and design of a project and on the construction process.

Real estate organizations, in addition to acquiring land, are frequently associated with developers as entrepreneurs and consequently may have an influential effect in generating building business.

In the foregoing vertical, functional classification of building industry elements, building trades labor as a separate component is not listed. This is because labor (as well as management) constitutes a horizontal element of the industry. It is obviously of major importance in the construction and materials manufacturing segments and, by its union work rules, may also affect the flexibility of choice available to the design professions. These considerations, however, are outside the scope of this book. The discussions in the chapters that follow are concerned with relationships between the owner and the architect-engineer and the owner and the construction contractor as entities, and do not explore the labor - management relations of either.

Categories of Construction

Construction activity may be classified from several different viewpoints, of which none are mutually exclusive. A custom of long standing differentiates between building construction and heavy construction. Heavy construction, sometimes called engineering construction, includes highways, bridges, railroads, airfields, dams, utilities, and similar works as well as certain types of industrial facilities. Many projects such as airports with their passenger and freight terminals, hangars, runways, taxi strips, ramps, etc., obviously share characteristics of both building construction and heavy construction; consequently a hard and fast line cannot always be drawn between the two. Nevertheless, the differentiation is real enough so that many construction firms confine their activities to only one of the two fields. The building industry therefore may be considered as a major component of the larger construction industry, using this term to embody both fields. This book is concerned primarily with contractual relationships involved in building design and construction activity, although these are not fundamentally different from analogous ones in the field of heavy construction.[1]

[1]For an analysis of contractual relationships with additional emphasis on heavy construction, see R.W. Abbett, *Engineering Contracts and Specifications* (New York: Wiley, 1963).

Another common breakdown of construction activity is that between private construction and public construction. Private construction includes all construction undertaken by private individuals, associations, corporations, or other business organizations, where financing is arranged through private channels. Public construction is initiated by an agency of the government—national, state, city, or other political unit—and is financed through special bond issues or by appropriations. All construction for or by the government is classified as public works. However, even in the area of building construction no firm line can be drawn between types of buildings sponsored by government enterprise and by private enterprise. Although building types such as post offices, jails, and courthouses are clearly within the purview of public construction exclusively, there are no major building types in which the provision is exclusively in the hands of private enterprise. Educational buildings and hospitals are readily recognized as areas of both private and public activity, and the diversity of facilities ranging from residential and religious to amusement and industrial, required in connection with the country's defense establishment, leaves few types that can be recognized as sponsored by private activity alone. Nevertheless, except in times of national emergency or severe economic depression, the volume of private construction has predominated.

A third classification divides all building activity into investment building and consumption building. In investment building monetary return is expected. It includes buildings erected for direct rental for any occupancy as well as those constructed by business organizations to house manufacturing processes or commercial activities (retailing, warehousing, amusement, etc.) where the monetary return through operating profits is indirect insofar as the building itself is concerned. Consumption building, on the other hand, includes the construction of all facilities from which no monetary return is expected, such as institutional and government buildings and most houses built by the owners for their own occupancy.

Finally, in connection with the compiling of building statistics and for purposes of economic analysis, the terms total construction and new construction are used. These terms permit a breakdown between new construction and that carried out as alterations, maintenance, and repairs. The importance of new construction is often not fully appreciated, especially because construction statistics normally report all structural additions and alterations to buildings as new construction, alloting minor alterations only to the maintenance and repair classification. Even on this basis, however, the long-range average indicates that roughly one-third of the total annual expenditure for all construction goes for maintenance and repair. The reason for a ratio of this magnitude becomes apparent when it is considered that new construction in any one year can create only a small addition to the total supply of buildings already in existence. Consequently, the bulk of activity that is carried on indoors must depend upon existing facilities. These require maintenance in order to continue to serve their original functions or alteration to meet the requirements of expanding needs and changed occupancy patterns.

Building Production as Manufacturing

The construction of buildings is an ancient art and some of its operational patterns that were established before the industrial revolution are still discernible in the industry today. Perhaps the two most significant of these are the organization of building labor into craft unions and the lingering concept of the building process as a hand-wrought field operation. However, growing awareness that this concept is of diminishing validity with respect to complex modern buildings has led to the prefabrication of ever-larger components, resulting in activities at the building site becoming more and more like an assembly operation. This is an indication that the building process is becoming industrialized and this trend cannot help but lead to changes in the patterns of building industry organization.

Nevertheless, the production of buildings differs from general industrial manufacturing in two significant respects. First, the immobility of the product requires that "final assembly" be accomplished at the place of ultimate use and frequently extensive preparation is required. This is in contrast to the manufacture of most commodities which are mass produced in a centrally located plant and shipped to the point of use. The second important difference is the method of financing production. In most industries, production is financed by entrepreneurial capital or short term bank loans and is independent of any financing required by the customer in purchasing the product. Building construction, on the other hand, is normally financed by mortgage loans which are made to the buyer (owner) rather than to the manufacturer (builder). Since the product itself constitutes the security for the loan, the lender is in a position to exert an influence on both design and construction operations as noted earlier.

Despite these two quite fundamental differences, building construction is essentially a manufacturing operation and is likely to become even more so. Therefore it is subject to the same factors of labor relations, management, and economics that characterize manufacturing in general. These often appear in different form, however, because of the industry's loose organization and its dependence on the voluntary co-operation of its diverse elements to achieve unified action.

Building as a Service Industry

The great bulk of building activity is generated by sources outside the building industry itself. In contrast with the procedures of many other manufacturing industries, only a relatively small portion of building activity is initiated by the builders in advance of a specific customer. Except for the operative builders in residential construction, building activity normally begins only after a buyer (owner) has determined his need for a new facility and has decided to procure it.

The design segment of the industry frequently learns of a proposed project first since architectural and engineering services are often enlisted in the formative stages. However, the pattern remains the same since planning and design do not get under way until a client appears. It is this predominance of external decision with respect to the initiation of activity, and consequent volume of work, that characterizes building as a service industry.

In addition to the operative home builders who produce houses for a market rather than for specific customers, the owner-builder companies are also exceptions to the service concept. These are construction organizations that initiate construction personally, especially in the fields of apartment and office buildings, with a view to subsequent sale or to operation as an investment property. A more recent development is the entry of large materials producers into this type of activity, especially with respect to extensive projects comprising several buildings embracing both residential and commercial occupancies.[2] In projects such as these, the services of private architects, engineers, and building contractors are utilized, but management control of both design and construction may be exercised by the sponsoring corporation's own agency created especially for the purpose.

Notwithstanding these exceptions and also allowing for the promotional activity carried on by all segments of the industry, the predominant character of building remains that of a service industry. In large measure, this will undoubtedly continue to be so, since the industry's mission is to produce facilities to house the diverse activities of our socio-economic complex, and the requirements for these activities must, in general, be determined by agencies external to the building industry itself.

Architectural and Engineering Services

The scope of architectural and engineering (A & E) services varies widely, depending upon the nature of the project, the planning and programming capabilities of the owner's organization, and with corporate clients or government agencies, the extent to which the owner carries out supervision and inspection functions while the project is under construction. In any event, the principal source of guidance with respect to building projects will be the professional services of an architect or architect-engineer. As far as design organizations are concerned, architect or architect-engineer will be considered substantially synonymous for the purposes of this book and denote a single design responsibility. Whether an architectural firm has engineering capability

---

[2]For example, the Century City development in Los Angeles, sponsored by the Aluminum Company of America, which covers 180 acres and consists of apartment and office buildings, a hotel, and a shopping center all designed and landscaped in accordance with a master plan.

"in house" or secures it through retaining consulting engineers, the basic services furnished include all normal structural, mechanical, and electrical engineering required in connection with a building project. This is true whether the firm calls itself "Doe & Doe, Architects" or "Doe & Doe, Architects & Engineers." Even in the latter, outside consulting engineers are often retained to supplement the in-house staff.

Basic Services as defined by the American Institute of Architects (and discussed further in Chapter 2) are considered to be rendered in five phases:

Schematic design phase
Design development phase
Construction documents phase
Bidding or negotiation phase
Construction phase

Although in the schematic design phase the scheme of the project is intended to be determined after ascertaining the owner's requirements, it frequently develops that the owner's requirements have not been formulated with sufficient precision to permit preparation of meaningful schematic design studies. The degree to which this may be true varies, of course, with different clients, and it is for this reason that the concept of *project analysis services* has developed. These are performed before starting the schematic design phase and may consist of any or all of the following:

Feasibility studies
Financial analysis
Location and site analysis
Operational programming
Building programming

Whether such services are carried out by the staff of the architectural firm, by consultants retained by the architect for the owner, or by consultants retained directly by the owner, it is obviously desirable that a firm program for design of the facility be established before schematic studies of possible solutions are started. At a minimum the building program should resolve the matters of space requirements and relationships, occupancy requirements, and budget limitations.

The design development phase of the basic services normally produces definitive design drawings, an outline specification, and an estimate of cost for checking against the budget. Working drawings and the project specification are produced during the third or construction document phase, necessary clearances from the building department and other controlling agencies obtained, and any required adjustments made in the estimated construction cost. In the fourth phase the architect-engineer assists the owner in setting up and conducting bidding and award procedures or in obtaining negotiated proposals for the work.

The fifth phase of basic services covers general administration of the construction contract. This function includes the checking of shop drawings and samples, interpretation of the drawings and specifications, review of change orders, approval of certificates for payments to the construction contractor, and making periodic visits to the site for ascertaining whether the work is proceeding in accordance with the construction contract documents. This phase of the basic services is less extensive on much of the work for the federal government where construction agencies such as the Army Corps of Engineers, the Naval Facilities Engineering Command (formerly Bureau of Yards and Docks), and the General Services Administration are in a position to furnish their own staffs for a large portion of this administrative and supervisory activity. The same situation is true for many state public works departments and for some large private corporations that carry on a more or less continuous building program. On the other hand, the services of the architect-engineer firm under this phase may be expanded to include construction management as agent of the owner. Such responsibility for coordination and general management of the construction operation may be exercised in connection with the "separate contracts" system or the system of "several contracts" described in the following article. When, in addition to the basic services, an architect-engineer firm performs construction management and project analysis functions, the term "expanded services" or "comprehensive services" is often used.

The Contract Method of Construction

Construction by contract is the predominant method of accomplishing building projects in the United States. The prevalence of the method no doubt accounts in large measure for the custom of designating the constructors of buildings as "contractors." The alternative to the contract method is construction by force account, where the owner maintains a building labor force on his own payroll, furnishes all materials and equipment, and exercises direct management of the work. Corporate owners and certain government establishments possessing extensive physical plants that require a force of building trades workmen for maintenance operations frequently undertake alterations or additions with their own forces. Such construction activity, however, constitutes only a minor fraction of total building construction. The activities of operative home builders possess some of the characteristics of the force-account method, but these firms are basically construction organizations and their operations are not usually considered construction by force account.

Traditionally, the contract method of construction has operated under the general contracting concept supplemented by the subcontracting procedure. The general contractors constitute top management in a large sector of the building industry, and it is they who assemble as a supplement to their own capabilities, the diverse combinations of specialty subcontractors required to construct a particular project. The amount of work performed by the general contractor's own forces will vary with the type of building and the character of his own organization, but a substantial portion will of necessity be subcontracted to those possessing the special skills required. In an effort to discourage the practice of "construction brokerage," some contracts limit the amount of work that a general contractor may subcontract.[3] However, with the ever-increasing complexity of modern buildings, subcontracting practice will probably expand rather than decrease. It is the existence of the large and diverse group of specialty subcontractors that makes possible the assembly of specialists by the general contractor for the project and their subsequent release at the termination of the work. This procedure provides the specialization within flexibility of organization that characterizes building construction operations.

The same technological factors that tend to accelerate the trend toward increased subcontracting have also operated to undermine the dominance of the general contracting concept. This is especially true in mechanical and electrical systems, where both the complexity of these divisions of the work and the proportion of the building budget represented by them have increased rapidly in recent years. This situation has given rise to the "separate contracts system" which is generally understood to connote that the owner awards separate contracts for general construction, heating and air conditioning, plumbing, and electrical work; heating and air conditioning and plumbing are frequently combined into one mechanical contract.

To the single general contract and the separate contracts systems may be added a third, sometimes called the system of "several contracts." Usually, this means the awarding by the owner of separate contracts for all phases of the project, thereby eliminating the general contractor entirely. Since the general management of the construction operation must be centralized somewhere, this procedure requires that the owner possess his own construction management organization which takes responsibility for coordinating and supervising the operation or that these functions be performed by an architect-engineer firm qualified to offer this type of service.

The Single Contract System

The predominant characteristic of the single contract system is centralized responsibility for management of the construction operation. The general contractor assumes full responsibility to the owner for completion of the entire project in accordance with the plans and specifications, including work carried

---

[3]For example, contracts of the Public Buildings Service of the General Services Administration limit the amount of work that may be subcontracted to 88% of the prime contract value.

out by subcontractors as well as that accomplished by his own forces. He is responsible for completing the building on time and at the agreed cost and assumes liability for the actions of his subcontractors. There is no contractual relationship between the owner and any subcontractor. This undivided responsibility characterizes the single contract system whether the construction contract has been awarded through competitive bidding or by negotiation.

Although the bulk of building construction outside the home building field continues to be accomplished by competitive-bid lump-sum contracts, the single contract system has been sharply criticized in recent years because of abuses centering on the procedures for awarding subcontracts. Foremost is the practice known as "bid shopping" in which after the general contract is awarded, the successful contractor endeavors to obtain subbids lower than those originally submitted. On the other side of the picture is the practice of "bid peddling" where the successful general contractor is offered lower prices by unsuccessful subcontract bidders than those he used in preparing his winning bid. These practices, if left unchecked, tend to drive out the more competent major subcontractors who prepare valid bids for quality work; where they are prevalent, they constitute additional factors to those discussed in the preceding article, tending to erode the dominant position of the single contract system.

Notwithstanding these inherent problems, there is increasing need in today's complex building operations for the coordinated construction management that the single contract system offers. It should also be borne in mind that the disabilities of the system are of differing significance with respect to private and public construction; and it does not appear that the matter of single versus separate contracts must necessarily be resolved on an "either/or" basis. Methods intended to achieve the merits of both systems in certain situations are discussed under bidding and award procedures in Chapter 7.

The Separate Contracts System

As noted earlier, the "separate contracts" of this system are generally understood to encompass separate mechanical and electrical contracts only in addition to a contract for the general construction work. Perhaps the term "multiple prime contracts" would be more closely descriptive since the mechanical and electrical contractors may subcontract portions of their work just as the contractor for general construction may subcontract his steel erection, roofing, etc. Throughout this book the term "separate contracts" will denote this connotation.

The growth of the separate contracts system is based on the fact that the mechanical and electrical systems of many types of modern buildings now account for approximately half the total project cost. In several types of facilities these environmental control systems are of greater significance to the owner than the structural fabric of the building. Consequently, a feeling has developed that a direct relationship between the owner and the contractors for this work should be established. The separate contracts system permits final selection of the mechanical and electrical contractors by the owner in consultation with the architect-engineer and establishes a direct channel of communication during progress of the work. This may prove to be of considerable advantage in setting up service arrangements for maintenance of the mechanical and electrical systems after completion of the building.

A few states have statutes requiring that all state buildings be erected under the separate contracts system; in most states, however, choice between single and separate contracts is left to administrative decision. A similar situation exists in the larger municipalities for their public works construction. Corporate owners with continuing building programs follow varying practices. In any event, it should be noted that additional services beyond the basic services will usually be required of the architect-engineer where separate contracts are employed.

Lower construction costs are claimed by both the proponents of separate contracts and by the advocates of the single contract system.[4] The claims of each group frequently appear to be documented by instances cited in their favor, but there is much uncertainty with respect to interpretation of data and comparability of premises assumed in the argument. Since the architect-engineer's primary responsibility is to the owner, his recommendations in this connection (when a choice exists) presumably will be based on his belief as to which system will be in the owner's best interest for the specific project. Many architectural and engineering firms find that their range of clients and projects requires that they be prepared to carry out construction contract administration functions under either system.

Unit Design – Construction Services

Although it is common practice in this country for building construction contractors to execute work on the basis of plans and specifications prepared outside their own organizations, a substantial amount of building, particularly in the industrial field, is accomplished under single contracts covering both design and construction. An important characteristic of this type of operation is the high degree of responsibility assumed by firms offering such services. This feature of combined services appeals to owners who desire to obtain new facilities with minimum involvement in the complex relationships that characterize the building industry.

[4]The Associated General Contractors of America are strong proponents of the single contract system, whereas the National Association of Plumbing-Heating-Cooling Contractors and other associations of mechanical speciality contractors just as strongly advocate the use of separate contracts.

Of the disadvantages cited against this practice the principal one appears to be the elimination of the architect-engineer as a disinterested professional advisor to the owner; that is, all design and construction considerations are in one package and there may be no impartial judge of construction performance. The concept of comprehensive architectural services mentioned earlier provides a mechanism whereby an owner may delegate responsibility to any degree desired without eroding the architect's unbiased position as the owner's agent. The cogency of arguments concerning the relative merits of unit design-construction contracts, comprehensive architectural services, and conventional practice procedures necessarily depends somewhat on the planning, programming, and construction management capabilities of the client's own organization.

Types of Construction Contracts

Building construction contracts fall into two broad categories based on the method of award: competitive-bid contracts and negotiated contracts. They are also classified with respect to basis of payment as lump-sum or cost-plus-fee contracts. Competitive-bid lump-sum contracts constitute the most widely used combination for building construction and are mandatory for almost all public works construction. Although there is no reason why lump-sum contracts cannot be negotiated as well as bid competitively, the cost-plus-fee types are more common in private work when award is made through direct owner-contractor negotiation. In fact, the use of negotiated contracts of any type is limited largely to private construction except in times of emergency when certain government agencies, especially those handling construction for the armed forces, are granted negotiating authority.

Lump-sum contracts, as the name implies, are based on a fixed price, agreed upon in advance, for which the contractor will perform all of the work required for construction of the project in accordance with the plans, specifications, and other contract documents, his profit being included in the lump sum. The basic principle underlying all forms of cost-plus-fee contracts is that the contractor will be reimbursed by the owner for all costs attributable to the project and will be paid, in addition, a fee for his services. The amount of the fee may be determined in several ways: as a percentage of the actual cost of construction; a fixed fee; a fixed fee with guaranteed ceiling; or a fixed fee with a profit-sharing arrangement in event the cost turns out to be less than the guaranteed ceiling. It should be noted here that cost-plus-percentage contracts have practically disappeared due in large part to the ease with which they can (and have been) abused. Under such contracts there is absolutely no incentive for the contractor to hold costs down, and it enables an unscrupulous contractor to increase the amount of his fee by inflating the cost of the work, thereby being rewarded for inefficient operations.

Aside from the basis of payment, lump-sum and cost-plus-fee contracts have many features in common. By and large the "Conditions of the Construction Contract" are relevant to both types and hence are discussed in Chapter 4 before the more detailed consideration of lump-sum and cost-plus-fee agreements in Chapters 8 and 9, respectively.

Industry Ethics and Standards

The wide influence of building activity on the country's economy and on the character of the physical environment that it produces obligates all segments of the industry to be aware of the responsibilities involved. Shoddy performance in building has long-lasting effects since the products of construction are so durable and expensive—even when defective—that they cannot readily be discarded. A poorly planned school will last as long as an efficient one, given the same quality of constructon; and a poorly built hospital probably will have to be used as long as a well built one, but at constantly increasing maintenance expense. Thus the technical sufficiency of design and construction has social and economic ramifications that are of paramount significance to the owner and the community.

It is necessary that a workable code of ethics regulate the professional and business relationships among the several elements of the industry. *The Standards of Professional Practice* of the American Institute of Architects (AIA Document J330) is representative of similar codes established by other professional components of the design professions. It is reproduced in full in Appendix A. The Associated General Contractors of America have promulgated a *Code of Ethical Conduct*,[5] and many other trade associations in both the construction and materials manufacturing segments of the industry issue standards which seek to assure fair operating procedures and consistency of quality in the production of building materials and components. It would be unrealistic, however, to assume that 100% compliance with the several standards is the rule. Nevertheless, pressure to eliminate questionable practices must be maintained so that the diverse elements of this loosely organized industry may achieve viable working relationships in their common interest of efficient building production.

Chronology of a Building Project

The origin of a project occurs when an owner decides to build or at least decides to explore the possibility of building. This decision is prompted by widely differing factors that depend on the type of facility, the need for it, and the earning potential for commercial, industrial, and investment enterprises. The principal stages and events in the course of a project's design and construction are identified for easy reference in the following concise overview of operations.

[5]See Appendix B.

FEASIBILITY AND SITE ANALYSIS

Regardless of building type, the questions of feasibility and site characteristics must be considered first, and it is highly important that architectural (and often engineering) services be available during these preliminary studies. Physical and environmental conditions at the site may exert strong influence on the type of structure required to house the contemplated functions; and when market potential and financial analyses are being carried out by economic consultants, the concurrent availability of architectural consultation will bring technical considerations to bear on the many factors that call for joint resolution.

PROGRAMMING

The feasibility and site studies lead directly to the programming stage where occupancy requirements and space relationships are coordinated with budget considerations and financing.

DESIGN

When the program has been approved by all concerned, the project is ready to proceed through the schematic and design development phases of basic architectural and engineering services.

WORKING DRAWINGS AND SPECIFICATIONS

These and other documents required for bidding and construction are prepared from the definitive design drawings and outline specification developed in the design phases.

BIDDING

If construction is to be accomplished by competitive-bid lump-sum contract, invitations to bid (or advertisements with public work) are issued, and bids are received on a specified date at the end of the bidding period.

AWARD OF CONSTRUCTION CONTRACT

After analyzing and checking the bids, the architect makes recommendations to the owner, who then proceeds to award the contract and authorize the start of construction.

CONSTRUCTION STAGE

During the progress of construction the architect-engineer carries out his contract administration functions as delineated in the agreement between owner and architect.

SUBSTANTIAL COMPLETION

This occurs when construction of the building is sufficiently completed so that the owner can take occupancy.

FINAL COMPLETION

This is reached after final inspection, final acceptance by the owner, and final payment.

Chapter 2

Contracts for Architectural and Engineering Services

Selection of the Design Firm

In the preceding general discussion of building industry relationships it was pointed out that basic design services customarily include all normal structural, mechanical, and electrical engineering required for a project and that this is true whether the design firm has in-house engineering capability or secures it by retaining consultants in these fields. Both organizational patterns are used effectively, and the prevalence of either in any particular section of the country seems largely a matter of custom, although influenced by the number of competent consulting engineering firms available locally. Some architectural firms with offices in different parts of the country operate with in-house engineering in one region but follow the consultant pattern in another. Either way, the owner has single design responsibility, a prime requisite if effective professional services are to be rendered.

Some independently practicing architects and consulting engineers form a temporary association or joint venture for purposes of obtaining and executing certain projects. Here the designation "Doe & Doe, Associated Architects and Engineers" is often employed. In such cases, however, it is necessary that *one* member be designated to represent the group officially in its dealing with the client, the actual allocation of responsibility within the group having been determined by mutual agreement before receiving the professional services contract.

Architect and architect-engineer firms should be selected on the basis of their qualifications and experience and their compensation should be established by negotiation. Competitive bidding is neither suitable nor appropriate for architectural and engineering work because close definition of the quality and extent of such services is not possible and, consequently, there is no realistic basis for comparison of proposals that might be tendered.[1] In public works construction the courts have sustained repeatedly the position that award of professional service contracts does not require competitive bidding.

The great majority of architect-engineer commissions are awarded by *direct selection* or *comparative selection.* Direct selection is most often employed by private individual clients, especially when undertaking projects of moderate scope. The comparative method is more often employed by corporate clients, building committees, and other groups or associations and is the predominant method of selection used by government agencies. Both procedures normally encompass three basic steps:

1. A review of the firm's qualifications and experience record, together with photographs and other documentation of executed work.

2. An interview with key members of the firm to give the investigator or selection committee an opportunity to assess the firm's attitudes, philosophy, and management capability.

3. An investigation of selected previous projects (preferably through visits to buildings) and conversations with former clients.

The comparative selection method is usually initiated by the owner extending invitations to submit an application, together with a statement concerning education and experience of key personnel, size and type of organization, representative projects executed by the firm, etc. Invitations may be limited to a restricted list of firms or public announcement may be made as frequently occurs for public buildings of special national or regional significance. After review of applications, a smaller selected group of firms is usually invited for personal interview. Final selection should be based on evaluation of a firm's general reputation within the design professions, its design competence, technical competence, integrity, managerial capacity, and manpower available for assignment to the project.

Adequacy of Agreements

It is in the interest of both owner and architect that a written agreement be executed delineating the relations and obligations between the parties. The complexity of modern building design and construction and the many unforeseen situations likely to arise during the course of a project render an oral

[1]Both the American Institute of Architects and the American Society of Civil Engineers (among other professional societies) consider price competition for professional services to be unprofessional conduct. See AIA Document J330 (in Appendix A); also *Guide to Professional Practice Under the Code of Ethics,* ASCE, June 1964.

agreement or exchange of letters inadequate for record purposes if there are misunderstandings and disputes. It is particularly important that the client understand the terms and conditions under which the architectural and engineering (A & E) services will be performed and the costs for which he will become obligated.

When a letter form of agreement appears to be called for in lieu of a more formal contract document, it should incorporate by reference an attached definitive statement of services to be rendered and compensation to be paid. If such a statement, however, is much less inclusive than the *Terms and Conditions of Agreement Between Owner and Architect* that constitutes a major portion of the AIA standard forms for owner-architect agreements (to be discussed later), it is not likely to achieve its purpose. The adequacy of the letter of agreement procedure in any specific situation should be checked with legal counsel.

Usually, the drafting of a proposed agreement is preceded by conferences between prospective client and architect, at which time the character and general scope of the contemplated professional services are established and the amount of the fee negotiated. The architect then embodies these conclusions in a formal written proposal which may take the form of a definitive contract signed and sent to the owner for his signature. This proposal may be modified through negotiation (or even withdrawn) any time before its acceptance. Once accepted by the owner however, it becomes a contract.

Types of A & E Contracts

There are five principal types of contracts for architectural and engineering services classified by basis of compensation. Regardless of the scope of services to be performed, payment for them may be made on any of the following bases:

Percentage of construction cost
Lump-sum fee
Multiple of direct personnel expense
Professional fee plus expense
Professional fee plus expense with guaranteed ceiling

Where project analysis services are to be performed in addition to basic services as discussed in Chapter 1, one of the last three methods listed is frequently combined with the first or second. This allows matching the form of agreement to the character of work performed in each of these stages of the project. The relative merits of the several types of contract and factors governing their selection are discussed below.

PERCENTAGE OF CONSTRUCTION COST

This type of contract has a long record of wide and satisfactory use for A & E services. It is the simplest type of agreement to operate and for the most part provides compensation to the architect and costs to the owner in direct proportion to work performed. It is particularly well adapted to projects of definite scope where bids are to be taken on complete plans and specifications before starting construction.

One disadvantage from the owner's viewpoint is that he does not know the precise amount of his obligation for professional services when he signs the contract. He may also feel that the architect may attempt to inflate the cost of the project in order to increase the fee. Any such procedure would, of course, be highly unprofessional and unethical, but in any event budget limitations on most building projects make this factor a not very real one.

This method has two possible disadvantages for the architect. When skill and effort in planning and design produce a solution to the owner's program that results in construction cost less than the amount originally budgeted, the fee is correspondingly reduced, thereby penalizing the architect for his efficiency. Furthermore, on very large projects where the fee percentage is normally lower than on smaller ones, an owner's extensive use of change orders to accomplish additional work can result in fee increments that are insufficient to cover the amount of architectural and engineering work required.

LUMP-SUM FEE

This type of contract obviously eliminates the two uncertainties noted that an owner may feel with respect to the open endedness of the percentage type based on construction cost. The lump-sum fee may be based on a percentage of the estimated construction cost or on the design firm's detailed analysis of estimated man-hour costs of performing the professional services.

The first method is predicated on the premise that, for any specific building type, the construction budget available is a measure of the scope of the project and hence of the scope of architectural and engineering work required. It is widely used in private work and formerly was also the basis of fee determination for most lump-sum A & E contracts relating to public works construction. In June of 1967, however, a directive was issued to the military construction agencies by the Department of Defense calling for the exclusive use of the detailed analysis method, and a similar directive was promulgated by the General Services Administration. It should be noted that the lump-sum basis of compensation is customary in U.S. Government contracts for architectural and engineering services, although other types are occasionally employed in special situations.

MULTIPLE OF DIRECT PERSONNEL EXPENSE

Under this type of contract compensation for services is determined by summing the cost of salaries (plus mandatory and customary employee benefits) of personnel directly engaged on the project and multiplying this amount by a suitable factor. Under current conditions this factor is ordinarily not less than two and a half or three and represents indirect overhead costs, net fee or profit, and sometimes the participation of the firm's principals. If the firm's principals are not accounted for in the multiplier, a specific statement in the agreement should stipulate a fixed rate for principals' time and their identification made. When the architect retains consulting engineering services, his agreements with consultants are normally of the same nature as the owner-architect agreement, with any costs due to coordinating the work of consultants appearing as a direct personnel expense of the architect.

This type of agreement is particularly useful where the scope of services required cannot be determined with a reasonable degree of precision, which may occur with feasibility studies or the programming phase of project analysis services. It is also appropriate in situations where only partial professional services are required. Obviously a contract of this type requires having the design firm's accounting system adequate and available for audit by the owner.

PROFESSIONAL FEE PLUS EXPENSE

There are two common forms of agreement under this type of contract. This title generally applies when compensation is a combination of fixed fee and a multiple of direct personnel expense. The designation *Cost Plus Fixed Fee* is often used when all of the architect-engineer's direct costs (personnel and other) and a portion of indirect overhead expenses are reimbursed by the owner. In both cases the fixed fee represents compensation for the professional skill, responsibility, and participation of the principal or principals and the availability of the resources of the firm. Reimbursement for expenses is simpler under the multiple of direct personnel method than where the more precise determination of all expenses is attempted. Here, extensive accounting records are necessary and considerable difficulty may be experienced in segregating costs when several projects are being handled in the A & E office concurrently.

This type of contract is often advantageous when it is possible to define the general scope of a project with sufficient precision to estimate the extent of direct involvement by the firm's principals but not possible to so establish the extent of technical services required at the time a contract is executed. The professional fee is then completely independent of final design costs unless the project scope is changed.

PROFESSIONAL FEE PLUS EXPENSE WITH GUARANTEED CEILING

When the open-ended character of the fee-plus-expense type contract is not acceptable to the owner and when a reasonably close approximation of the entire project scope is possible, the provision of a guaranteed ceiling for total cost may be incorporated. A distinguishing feature of this arrangement as opposed to lump-sum agreements is that the owner receives any savings that may result from actual expenses being less than originally estimated. However, unless the project is of such character that architectural and engineering costs can be closely estimated in advance, it is only prudent that a suitable contingency allowance be included when establishing the guaranteed ceiling.

Determination of Fee

From a strictly cost accounting point of view determination of fees for architectural and engineering services as a percentage of the construction cost may appear to be lacking a sound and practical basis. Nevertheless, its wide usage over the years has led to developing some approximate correlations between cost of construction and cost of performing related A & E services. Even when lump-sum fees are established by estimating detailed man-hour costs or where estimates of total costs of services are prepared under any other type of contract discussed in the preceding article, these are usually checked for reasonableness against recognized percentage fee schedules. Obviously no standard schedule of rates for performing professional services can be inclusive enough to provide for the many variations among building projects of even the same general character. However, the recommended schedules that have been promulgated from time to time by groups within the design professions are based on wide experience in a variety of building types and may be used with confidence as a basis for the negotiation of A & E contracts.

Many chapters of the American Institute of Architects as well as state and regional organizations issue recommended schedules of appropriate fees for basic services. Although these vary in different parts of the country, the recommended percentage rates generally recognize the degrees of complexity among different broad categories of occupancy or building type. For example, the rate applying to a proposed one million dollar hospital in a certain region might be 7 to 8%, whereas that for a one million dollar warehouse might be 5 to 6%. The recommended rates also vary with the "dollar scope" of projects within the same category of complexity, especially when repetitive elements occur as in the typical floors of high-rise buildings and in low buildings of extensive ground area where the same structural-mechanical module is repeated often.

Both percentage and lump-sum fees will, of course, vary with the extent of services to be performed. The fee schedules are normally predicated on the full performance of the basic architectural and engineering services identified earlier. Abridgement of these services or the addition of project analysis or construction management services requires adjustment of the recommended rates.

In some schedules the amount of the recommended fee is based specifically on preparation of the working drawings, specifications, and other contract documents for awarding a single lump-sum construction contract; when the project is to be built under the separate contracts system or on a cost-plus-fee construction contract the amount of the fee is higher. This reflects the additional work required of the architect-engineer under these two procedures, particularly in the area of construction contract administration.

Finally, it should be noted that all such schedules are *recommended* fees only and are intended to represent minimum compensation consistent with the cost of rendering competent professional services.

Scope of Professional Services

It is of utmost importance to both owner and architect that the services to be performed should be clearly defined. This is particularly essential in agreements where the fee is determined as a percentage of construction cost or is a lump sum, since there must be a complete meeting of minds as to precisely what work will be accomplished under the basic services and what will constitute "additional services." In contracts where the architect's compensation is based on a multiple of direct personnel expense, the distinction between basic services and additional services is not significant. However, even here the services should be divided into phases as discussed in the preceding chapter so that the owner may exercise control in the approval of planning and design at each phase.

The discussion of scope of professional services presented in the next three articles is keyed to contracts of the lump-sum or percentage-of-construction-cost types, but the descriptions of the functions to be performed are equally applicable to other types of agreements regardless of the basis of compensation. Although the discussion is independent of any standard contract form, two such forms have been included at the end of this chapter for reference.[2] It will be profitable for the reader to correlate the various topics covered with corresponding provisions in these standard documents. Although no direct reference to article numbers is made, a brief study of the documents will reveal the pattern of their structure and facilitate cross reference.

Because the scope of architectural and engineering services during the construction phase of a project is subject to considerable variation that is dependent upon the owner's in-house construction management capability, the scope of services rendered prior to award of the construction contract and those performed subsequent to award will be discussed separately.

Services Prior to Award of Construction Contract

Aside from programming or other project analysis services discussed in Chapter 1, the scope of work before the construction contract is awarded is that

---

[2] American Institute of Architects Document B131; and Form NavFac 4-4330, the standard U.S. Government form as adapted by the Naval Facilities Engineering Command.

contemplated in the first four phases of the basic services reidentified here for convenience:

Schematic design phase
Design development phase
Construction document phase
Bidding or negotiation phase

Since an owner signs a contract for A & E services in order to procure the complete design of a project together with all necessary drawings, specifications, and other documents required for its construction, the scope article of the agreement is often worded very broadly. This may take a form such as "The professional services contractor shall provide architectural and engineering services of every kind required in connection with the studies, designs, and the preparation of plans and specifications for this project." When an owner insists on this kind of statement, it is extremely important that the architect make certain that all possible professional services required for the particular building type be identified at the time the fee is established. Otherwise, he may be liable for providing specialists such as food handling and kitchen equipment consultants, and others, who are frequently considered to be outside the scope of basic services. A clause such as "... except as otherwise provided herein," appended to the general statement is helpful in such cases since it does provide for a listing of agreed-upon exceptions.

It will be noted that the standard form of agreement recommended by the American Institute of Architects as AIA Document B131 approaches this situation differently. The general statement requires the architect to "... provide professional services for the Project in accordance with the Terms and Conditions of this Agreement." The statement of "Terms and Conditions" then designates in detail just what is included and is not included under the contract. Protection of both owner's and architect's interests may be achieved either way as long as there is agreement regarding the scope.

DESIGN PHASES

Although all agreements are not organized into the same phases as those already identified, a review of contract forms discloses that some type of preliminary design phase is recognized as distinct from the preparation of working drawings and specifications. During this phase preliminary drawings and descriptive outline specifications are prepared and submitted to the owner. These are revised as necessary until a preliminary scheme or schematic design study is approved. The degree of detail incorporated in drawings, outline specifications, and cost estimates at this stage depends on whether a subsequent design development phase is called for in the contract or whether the "preliminaries" have been sufficiently developed so that working drawings and specifications can be started as soon as approval is given by the owner.

In general, recognition of both schematic and design development phases is greatly to the advantage of both owner and architect. This permits flexibility in studying solutions to the functional program in relation to alternative schemes for structural and mechanical systems without pursuing more detailed engineering design until a generally satisfactory overall scheme is approved. Then during design development phase the principal quantitative engineering problems can be worked out and problems of coordination between functional requirements and engineering systems resolved. With this procedure the production of working drawings and specifications can proceed smoothly, since any changes required in the building program presumably will have been made before the owner's approval of the definitive design documents.

CONSTRUCTION DOCUMENTS AND BIDDING

Once working drawings and the detailed engineering design required for their completion are under way, changes from the previously approved design development documents are expensive to make. Consequently, A & E contracts customarily provide that the owner shall bear the cost of such changes unless they are initiated by the architect.

All documents for bidding purposes are usually prepared under the construction document phase or, for government work, standard forms for instructions to bidders, proposal submittal, etc., are supplied by the agency contracting for the project. Government construction agencies for the most part conduct their own bidding operations, whereas in private work the bidding or negotiation phase customarily is managed by the architect on behalf of the owner. Bidding and award procedures are discussed in more detail in Chapter 7.

Services Subsequent to Award of Construction Contract

ABRIDGED CONSTRUCTION PHASE SERVICES

As noted earlier the scope of architectural and engineering services provided during the construction phase will depend on whether the owner maintains construction management capability within his own organization. United States Government construction agencies customarily possess this capability; consequently, A & E services after award of the construction contract are usually limited to the checking of shop drawings, material samples, and other submissions by the construction contractor, and to consultation that may be necessary to amplify or clarify the intent of plans and specifications. All other construction contract administration functions are carried out by Government personnel. Similar conditions sometimes prevail in private work where corporations that have continuing building programs maintain their own construction management and inspection forces. The situation is quite different, however, with respect to most private clients and institutional work such as schools, hospitals, and much college and university building. Here complete basic services are the rule and the architect acts as the representative of the owner throughout the construction phase.

FULL CONSTRUCTION PHASE SERVICES

Although the architect-engineer is not a party to the construction contract between owner and contractor, his status and functions as representative of the owner during construction are delineated therein. These administrative functions (usually stated in the General Conditions of the construction contract) are made a part of the owner-architect contract by reference; they will be considered further in Chapter 4. Discussion here will be focused on key provisions relating to scope and responsibility that should always be incorporated directly in the contract for architectural and engineering services.

The first point in which clear understanding is mandatory concerns the premise that the architect-engineer does not *guarantee* the work of the construction contractor. Since he is not a specialist in construction methods, techniques, sequences, or procedures (the sphere of presumed special competence of the building constructors), the architect cannot assume responsibility for the construction operations. Neither can he be responsible for the contractor's failure to carry out the work in accordance with the plans and specifications. Although in carrying out his construction contract administration functions, the architect-engineer makes visits to the site to ascertain whether work is proceeding in accordance with the construction contract documents and although he may take certain necessary corrective measures, the *responsibility* for proper execution of the work rests with the construction contractor. It is only the construction contractor who has made a contract with the owner for execution of the work.

SUPERVISION OF CONSTRUCTION

Although there is general understanding throughout the building industry of the spheres of responsibility assigned by custom to the design and construction segments when a project is under construction, people outside the industry are often confused on this point. Much of this confusion can be attributed to the special connotations given to commonly used words when employed within the industry. Probably the most troublesome term is "supervision" which is frequently confused with "superintendence." Within the industry the supervision furnished by the architect and superintendence furnished by the building contractor are understood to be two quite distinct functions. However, in general usage superintendence and supervision tend to become synonymous along with the closely related terms management and direction. It is understandable therefore that laymen, clients, and sometimes the courts interpret these terms differently from professionals in the building field, with the result that architects and engineers sometimes find themselves charged with responsibilities they never intended to assume. Reference to "supervision of construction" has led to interpretations by the courts that have held architects and engineers liable for acts of building contractors over which they had no control whatever.

In order to reduce the risk of various interpretations when disputes arise, the precise meaning of terms such as supervision and administration should be clearly defined in both the architectural and engineering contract and in the construction contract articles that describe the status and functions of the architect. The American Institute of Architects in the 1966 Revision of its Standard Documents[3] eliminated the word supervision entirely in an effort to clarify the scope of the architect's responsibility. The purpose of his on-site observations is to ". . endeavor to guard the Owner against defects and deficiencies in the work of the Contractor;" not to *supervise* his work.

PAYMENT CERTIFICATES

In this important area of contract administration it is also necessary that owner and architect have a mutual understanding concerning the extent of the architect's responsibility as provided in the professional services contract. When the project is built under a lump-sum construction contract, payments to the contractor are usually based on estimates of amount of work completed to date as related to a *schedule of values* for different portions or components of the project prepared immediately after award.

As a result of the architect's on-site observations and review of the contractor's application for payment (together with its supporting data) he issues a *certificate for payment* to the owner. This certificate represents that construction has progressed to the point indicated and that to the best of the architect's knowledge and belief it is in accordance with the drawings, specifications, and other construction contract documents. In many construction contracts the contractor is obligated to support the items of his application for payment ". . by such substantiating data as the Owner or Architect may require." The character and extent of such supporting information should be designated in the owner-architect agreement so that in event of future dispute the architect will not be vulnerable to charges that payment was certified on the basis of insufficient evidence. If auditing of the contractor's records is contemplated, it must be clear who is to pay for such services.

CHECKING OF SHOP DRAWINGS

Provision for this function is incorporated in all A & E contracts whether or not construction phase services are abridged. Shop drawings include all drawings, diagrams, performance charts, and other information prepared by the contractor, any subcontractor, or manufacturer to illustrate some portion of the project. Approval of shop drawings by the architect involves an element of risk

[3]AIA Document B131 at the end of this chapter and AIA Document A201-General Conditions of the Contract for Construction, reproduced at the end of Chapter 4.

when mistakes appear later. Because of this the extent of responsibility assumed in the approval of shop drawings is stated precisely in the owner-architect agreement and in the General Conditions of the construction contract. The pertinent provisions should indicate specifically that checking and approval of shop drawings relate only to conformance with the design concept and overall compliance with the plans and specifications. Such approval should never imply responsibility for their accuracy, especially in matters of fabrication or construction techniques which are outside the professional competency of the architect-engineer.

Furthermore, such approvals do not cover changes from the working drawings and specifications unless the contractor has called attention specifically to such changes. The same provisions apply generally to the approval by the architect of physical samples of materials and components which are intended to illustrate color, texture, workmanship, performance characteristics, etc. In other words, approval of shop drawings and samples does not relieve the construction contractor of his obligation to perform the work in accordance with the drawings and specifications on which the award was made.

Services Not Included

As noted under the discussion of services performed prior to award of the construction contract, provision should be made to identify specific items not included in the basic services. These fall under the two general headings of information to be furnished by the owner and items considered as supplementary or additional services for which the architect will be reimbursed separately.

OWNER-FURNISHED INFORMATION

It is generally accepted that the owner should furnish a certified survey of the site of the project and that the architect is entitled to rely on its accuracy. This survey should encompass zoning and deed restrictions, easements, rights of way, and encroachments; grades and lines of streets, alleys, sidewalks, and adjoining properties; boundaries and contours of the site; locations and dimensions of existing buildings, other improvements, and trees; and information concerning service and utility lines. Although practice varies, a common procedure for obtaining the survey is for the owner to contract directly with a firm specializing in civil engineering and land surveying.

It is also common practice for the owner to furnish information concerning subsoil conditions. This usually entails the services of a soils engineer who may obtain his data by means of test borings, test pits, and sometimes test piles. Here again practice varies in that the owner may either contract directly with the soils engineer or request the architect to retain consulting soils engineering services to be paid for as an additional service.

Of course, any contract for architectural and engineering services may be written to include in its scope any services on which owner and architect may agree. However, even in those contracts that call for "... all necessary architectural and engineering services of every kind required... ," it is necessary that both owner and architect have the same conception of just what is required. In addition, there are always changes that become necessary as a project develops and it is only equitable that some provision be made to cover their cost when they require revision of plans and specifications prepared on the basis of previously approved studies. The article on additional services contained in AIA Document B131 has been developed from wide experience and its provisions are equitable to both architect and owner. Many of the services listed may, of course, be incorporated in the scope of basic services if desired; study of the article, however, will disclose others which should remain in a list of exceptions to even the all-inclusive clause quoted above.

Among the items given which may readily be transferred to expanded basic services are those pertaining to project analysis services, the providing of detailed estimates, and the furnishing of interior design services. The remaining items, however, should stand as not included in the basic services.

Construction Cost

It is important in all types of A & E contracts that there be an unequivocal definition of this term, especially so where the fee is determined as a percentage of construction cost or as a lump-sum based on a percentage of the *estimated* construction cost. As used in this book the term construction cost is synonomous with the widely accepted concept of *net construction cost* that is, the cost of all work designed or specified by the architect-engineer but not including the cost of architectural and engineering services nor that of other planning and design consultants. Neither does net construction cost include the cost of land, rights of way, nor demolition of existing structures required to clear the site.

For purposes of fee determination the cost of work "designed or specified" includes all labor and materials. When labor or material is furnished by the owner, their cost is computed at current market values. The term materials is generally understood to include the fixed equipment required for the mechanical and electrical systems of the building. However, when costly items of large special equipment, both designed and procured by the owner, are to be housed, modification of this convention will be required. This situation may arise when the cost of the building is small in relation to the cost of the special equipment; certain nuclear facilities and many industrial projects fall into this category.

When a building is carried to completion, construction cost is simply the total cost of all work in place. In the event that construction does not proceed, however, it is necessary to have a method of determining its probable value for contract purposes. The most common bases for making such a determination are (a) the lowest reliable bid received or, if bids are not taken, then (b) the latest detailed cost estimate, or (c) the architect's last statement of probable construction cost when a detailed cost estimate has not been made.

Estimates and Limits of Cost

Very few building projects are developed without a ceiling placed on construction cost. This may be a fixed limit of cost known from the beginning, as with public buildings financed from appropriations or bond issues, or the the ceiling may be established during design development when the scope of the owner's program as well as the quality of construction, equipment, and finish are considered relative to the amount of money he is willing to spend. In this case estimates of probable construction cost are prepared from time to time and these function as an economic control in balancing the three interdependent factors of scope, quality, and cost. In both instances there is a high premium on accuracy of estimates made by the architect-engineer.

It must be recognized that a high degree of precision in these prebidding cost determinations is very difficult to achieve. This is true because market and competitive conditions in the industry *at the time of bidding* are dominant factors in establishing the level and range of proposals received. It is for these reasons that a bidding contingency should be included whenever a fixed limit of cost is established as a condition of the professional services contract or when such a limit is implied by a final estimate agreed on by owner and architect before calling for bids. Although the amount allowed for the bidding contingency will vary with circumstances, it rarely should be less than 10% of the limit or estimate.

It is also necessary that the extent of the architect's responsibility when no bids are received within the limit of cost be specified. A provision that has long been incorporated in certain U.S. Government contracts requires the architect-engineer, at his own expense, to revise the plans and specifications so that a usable facility may be built within the cost limitation. When such provision is made in private work, it should be stipulated that such services are the limit of the architect's responsibility in this connection and that on completion of this objective he will be entitled to the fees established in the professional services contract.

Payments and Reimbursable Expenses

BASIC FEE PAYMENTS

In lump-sum and percentage-of-construction-cost types of A & E contracts, provision is commonly made for payments at the ends of each phase of the basic services in accordance with an agreed upon schedule. Frequently, arrangements for a retainer upon execution of the contract and monthly payments as the work progresses are also included; these are so controlled, however, that total monthly payments at the ends of the phases will not exceed those established in the schedule. A widely used schedule recommended by the American Institute of Architects gives the following percentages of the total basic services fee due at the end of each phase:

Schematic design phase	15%
Design development phase	35%
Construction documents phase	75%
Bidding or negotiation phase	80%
Construction phase	100%

Under the multiple-of-direct-personnel-expense type contract, payments are made to the architect on a monthly basis as services are performed and attendant direct personnel expense incurred. Where a fee-plus-expense type agreement is used, payments are also made monthly as direct personnel expense is incurred, but payments on the fixed fee are limited in accordance with a schedule similar to the one previously mentioned.

PAYMENTS FOR ADDITIONAL SERVICES

The lump-sum and percentage types of contracts require further that the method of payment for additional services be stipulated. The most usual arrangement is to follow the procedure employed in multiple-of-direct-personnel-expense type agreements. It is frequently convenient, however, to carry out certain additional services under a lump-sum supplementary agreement or under a supplementary fee-plus-expense type agreement. Regardless of the method used, it is desirable to provide for segregation of billings covering services in addition to those relating to the basic fee.

REIMBURSABLE EXPENSES

The cost of travel and subsistence in connection with a project, as well as telegrams, long distance telephone calls, reproduction of plans and specifications for bidding purposes, fees paid to various governmental review agencies for plan check and approval, and similar items have, by custom, been considered as directly reimbursable expenses for which allowance is not made in the basic fee. In addition, some agreements place procurement of rendered perspectives and models in this category, as well as services of special consultants for other than normal structural, mechanical, and electrical engineering services. It is of utmost importance that there be a clear understanding on this point, since the scope of engineering services considered "normal" for one building type may be different in another.

Government contracts almost invariably require prior authorization for travel that is to be reimbursed separately, and some private owners set a limit to the amount of travel they will pay for as reimbursable expense. It is important that the various items allowed under this heading should be stipulated very carefully, since they can become a source of misunderstanding. Provision is usually made for payment of reimbursable expenses monthly upon presentation of a statement of expenses incurred.

General Terms and Conditions of A & E Contracts

These additional provisions should form a part of any contract for architectural and engineering services regardless of the type of agreement: ownership of plans and specifications, termination of contract, settlement of disputes, successors and assigns, extent and accessibility of accounting records. In addition, many special provisions are contained in government contracts because of statutory requirements. It is recommended that the reader compare articles covering these general provisions in the contract forms reproduced at the end of this chapter. Discussion of selected items follows.

Ownership of Plans and Specifications

It has long been a principle of professional practice that ownership of plans and specifications for a project resides with the architect or engineer who prepared them. Unless specific provision is made in the A & E contract however, it is doubtful whether the courts will uphold this point of view if the owner contests it. For this reason contracts for architectural and engineering services frequently define the drawings and specifications as "instruments of service" and state that they are the property of the architect-engineer and that they may not be used on other work except by agreement and with appropriate compensation. This provision is usually stated so that it will apply whether or not the project for which the documents were made is actually built. On the other hand, professional services contracts made by agencies of the U.S. Government require a clause stating that all designs and documents produced thereunder are the property of the Government and may be used on other Government work without additional compensation.

In private work the significance of the document ownership provisions will vary with the type of project. When multiple use of plans and specifications is contemplated from the beginning, this may be reflected in the original fee determination by establishing re-use "royalties" or similar devices. On public works projects the character of much of the work makes these provisions of little importance to the architect, except where his drawings are adopted as prototype designs for similar facilities that may be constructed at various locations. This possibility is usually known at the time the professional services contract is being negotiated.

Termination of Contract

Provision should always be made for termination because of nonperformance by either party and also for reasons beyond either party's control. Nonperformance by the owner might consist of failure to make payments promptly or not furnishing necessary information as agreed, and unreasonable delay or lack of technical competence would constitute nonperformance by the architect-engineer. Government contracts provide that termination may be made at any time either for the Government's convenience or the default of the professional services contractor. In private work a situation analogous to "convenience" may arise when the owner has to abandon a project due to failure to obtain financing, failure to obtain anticipated zoning changes, and similar reasons.

The contract should specify the method of settlement in event of termination, including procedures for notifying the other party of intent to terminate and establishing the basis of payment to the architect-engineer for work performed to date. It should be stipulated that, if termination is not the fault of the architect-engineer, he shall be paid his fee covering basic services performed to date of termination, charges for additional services and reimbursable expenses then due, and terminal expenses such as attorney's fees and those of other consultants required to establish the extent of work performed.

Settlement of Disputes

The best insurance against disputes arising under contracts for architectural and engineering services is an adequate written agreement, supplemented by carefully prepared plans and specifications. Nevertheless, even under the most favorable circumstances disputes must be anticipated and it is common practice to provide mechanisms for handling them short of litigation. In private work and in government work at some levels, provision for arbitration may be incorporated in the contract. Federal government contracts, however, prescribe a different procedure involving a decision by the contracting officer.

ARBITRATION

Arbitration is the voluntary submission of a dispute to an impartial board or *tribunal* made up of persons acquainted with the nature of the dispute. The method offers advantages of speed and economy over formal court proceedings and yet is sufficiently orderly to promote the reaching of reasonable decisions and awards. Contract provisions for arbitration of future disputes are enforceable under the laws of many states and under the United States Arbitration Act which would apply to a construction contract involving interstate commerce. Some state laws do not enforce such provisions, but they permit the parties to submit their dispute voluntarily to arbitration. When the parties so participate in the arbitration, judgment on the award may be entered in a court of proper jurisdiction and enforced as any other judgment. Also, when the applicable law permits enforceability of the arbitration clause but only one party appears at the hearing (ex parte hearing), the award is valid and may be confirmed in the same way. Of course, if the applicable law does not enforce the arbitration clause, the respondent is not obliged to participate and an ex parte award would not be confirmed.

If arbitration provisions are to have the effectiveness intended, it is obviously of prime importance that pertinent articles of the contract be drawn consonant with the laws of the state that will govern the contract. Some of these statutes prescribe rules and conditions governing arbitration procedures that must be strictly complied with if decisions and awards of the arbitrators are to be enforceable. The AIA owner-architect agreement form designated Document B131 provides that disputes arising thereunder shall be decided by arbitration in accordance with the Construction Industry Arbitration Rules of the American Arbitration Association. These rules have been so drawn that arbitration proceedings instigated under them will comply with applicable state statues.[4] Further discussion of arbitration procedures including selection of arbitrators, conduct of hearings, etc., is contained in Chapter 4, and the Construction Industry Arbitration Rules of the AAA appear in Appendix C for ready reference.

DECISION BY CONTRACTING OFFICER

The standard forms for professional services contracts used by agencies of the U.S. Government contain a clause stating that disputes concerning a question of fact ". . . shall be decided by the Contracting Officer. . . ." This term refers to the person executing the contract on behalf of the Government. His actual position varies within the organizational structure, but is usually at such level that his decision with respect to a dispute under the contract can be appealed to the secretary of the department or administrator of the executive agency concerned. However, since the contracting officer and his chief are both representatives of the government, the "owner" here is both a party to the contract and the judge of its performance.

Usually, disputes that cannot be disposed of by agreement are resolved by the contracting officer, but those involving large amounts of money such as claims for extra services due to (alleged) changes in scope are often appealed by the mechanism provided in the disputes clause. The decision of ". . . the Secretary . . . shall be final and conclusive unless the same is fraudulant or capricious or arbitrary or so grossly erroneous as necessarily to imply bad faith or is not

---

[4]However, the actual wording of the article on arbitration (Art. 11) in Document B131 should be checked with legal counsel to ascertain whether modification is required in any specific situation.

supported by substantial evidence." These exceptions provide the basis on which decisions may be subjected to judicial review as a last resort by the A & E contractor.

Professional Responsibility and Liability

Architects and engineers, like other professionals, are expected to exercise reasonable care and skill in carrying out their work. Although this does not imply that perfect plans and specifications will always be produced, the level of performance is required to be consistent with that ordinarily provided by other qualified practitioners under similar circumstances. In recent years, however, the area of law concerned with professional responsibility and liability has become very active and new or modified principles have developed that expand substantially the zone of risk in professional practice. One of the most significant relates to the liability of the architect-engineer to third parties unconnected with the contract for claims of negligence or errors in design that lead to alleged injury of persons using the building.

Other court decisions have led to extension of architect-engineer liability in the area of building product performance, that is, holding him responsible for damages caused by faulty materials and components and sometimes for the cost of their replacement. Obviously, tendencies in this direction place a premium on the selection and specification of building products with long records of satisfactory performance and inhibit introducing new materials and methods. In an effort to direct a larger share of such responsibility toward the manufacturer or producer of new products, the AIA has issued a policy statement on building product development and uses. This document which sets up certain obligations for manufacturer, architect, building contractor, and owner is reproduced in Appendix D.

Of course, the A & E contractor will be held responsible for errors and omissions in the plans and specifications and may be held negligent in preparing cost estimates when the lowest bid on a project exceeds a stated limit of cost. These are risks that cannot be avoided, but can be minimized by maintaining careful checking and administrative procedures throughtout the design phases.

During the construction phase, errors and omissions evolve from contract administration functions and may result in claims based on negligence in certifying payments to the contractor or on faulty "supervision" of the building contractor's work. It is for these reasons, as discussed earlier in this chapter, that the architectural and engineering contract must provide clear definition of functions and responsibility assumed during the construction phase.

Professional Liability Insurance

Although increased attention to quality control of drawings and specifications and careful indication of the scope of services in the contract documents can accomplish much toward reducing potential liabilities, the circumstances of contemporary practice make it prudent to carry professional liability insurance. The insurance policy commended to its members by the American Institute of Architects covers liability for errors, omissions, or negligent acts arising from the performance of professional services and the cost of legal defense including investigation, lawsuits, and arbitration. Provision of legal defense is of no little significance since the costs of defense, even in groundless claims, can involve very large amounts of money.

EXAMPLES OF CONTRACT FORMS

Two standard forms of contracts for architectural and engineering services given here are for illustrative purposes. The first, reproduced by permission of the American Institute of Architects, represents a percentage-of-construction-cost type widely used in private work; the second is the standard U.S. Government form as adapted by the Naval Facilities Engineering Command. A study of this form (which is used with GSA Standard Form 26 as cover) will disclose the many special items required in government contracts by federal statutes or executive orders. It is only rarely that a standard contract form can be used without some modification or amplification relating to the specific project in hand, and provision for the necessary addition of articles is usually contained in the form itself.

THE AMERICAN INSTITUTE OF ARCHITECTS

AIA Document B131

Standard Form of Agreement Between Owner and Architect

on a basis of a

PERCENTAGE OF CONSTRUCTION COST

AGREEMENT

made this   day of   in the year of Nineteen
Hundred and

BETWEEN

the Owner, and

the Architect.

It is the intention of the Owner to

hereinafter referred to as the Project.

The Owner and the Architect agree as set forth below.

This document is copyrighted by The American Institute of Architects and is reproduced here with its permission.

I.   THE ARCHITECT shall provide professional services for the Project in accordance with the Terms and Conditions of this Agreement.

II. THE OWNER shall compensate the Architect, in accordance with the Terms and Conditions of this Agreement, as follows:

a. *FOR THE ARCHITECT'S BASIC SERVICES,* as described in Paragraph 1.1, a Basic Fee computed at the following percentages of the Construction Cost, as defined in Article 3, for portions of the Project to be awarded under

A Single Stipulated Sum Contract	per cent(	%)
Separate Stipulated Sum Contracts	per cent(	%)
A Single Cost Plus Fee Contract	per cent(	%)
Separate Cost Plus Fee Contracts	per cent(	%)

b. *FOR THE ARCHITECT'S ADDITIONAL SERVICES,* as described in Paragraph 1.3, a fee computed as follows:

Principals' time at the fixed rate of   dollars ($)
per hour. For the purposes of this
Agreement, the Principals are:

Employees' time computed at a multiple of   (   )
times the employees' Direct Personnel Expense
as defined in Article 4.

Additional services of professional consultants engaged for the normal structural, mechanical and electrical engineering services at a multiple of   (   ) times the amount billed to the Architect for such additional services.

c. *FOR THE ARCHITECT'S REIMBURSABLE EXPENSES,* amounts expended as defined in Article 5.

d. *THE TIMES AND FURTHER CONDITIONS OF PAYMENT* shall be as described in Article 6.

TERMS AND CONDITIONS OF AGREEMENT
BETWEEN OWNER AND ARCHITECT

ARTICLE 1

ARCHITECT'S SERVICES

1.1   Basic Services

The Architect's Basic Services consist of the five phases described below and include normal structural, mechanical and electrical engineering services.

Schematic Design Phase

1.1.1   The Architect shall consult with the Owner to ascertain the requirements of the Project and shall confirm such requirements to the Owner.

1.1.2   The Architect shall prepare Schematic Design Studies consisting of drawings and other documents illustrating the scale and relationship of Project components for approval by the Owner.

1.1.3   The Architect shall submit to the Owner a Statement of Probable Construction Cost based on current area, volume or other unit costs.

Design Development Phase

1.1.4   The Architect shall prepare from the approved Schematic Design Studies, for approval by the Owner, the Design Development Documents consisting of drawings and other documents to fix and describe the size and character of the entire Project as to structural, mechanical and electrical systems, materials and such other essentials as may be appropriate.

1.1.5   The Architect shall submit to the Owner a further Statement of Probable Construction Cost.

Construction Documents Phase

1.1.6   The Architect shall prepare from the approved Design Development Documents, for approval by the Owner, Working Drawings and Specifications setting forth in detail the requirements for the construction of the entire Project including the necessary bidding information, and shall assist in the preparation of bidding forms, the Conditions of the Contract, and the form of Agreement between the Owner and the Contractor.

1.1.7   The Architect shall advise the Owner of any adjustments to previous Statements of Probable Construction Cost indicated by changes in requirements or general market conditions.

1.1.8   The Architect shall assist the Owner in filing the required documents for the approval of governmental authorities having jurisdiction over the Project.

Bidding or Negotiation Phase

1.1.9 The Architect, following the Owner's approval of the Construction Documents and of the latest Statement of Probable Construction Cost, shall assist the Owner in obtaining bids or negotiated proposals, and in awarding and preparing construction contracts.

Construction Phase—Administration
Of The Construction Contract

1.1.10 The Construction Phase will commence with the award of the Construction Contract and will terminate when final payment is made by the Owner to the Contractor.

1.1.11 The Architect shall provide Administration of the Construction Contract as set forth in Articles 1 through 14 inclusive of the latest edition of AIA Document A201, General Conditions of the Contract for Construction, and the extent of his duties and responsibilities and the limitations of his authority as assigned thereunder shall not be modified without his written consent.

1.1.12 The Architect, as the representative of the Owner during the Construction Phase, shall advise and consult with the Owner and all of the Owner's instructions to the Contractor shall be issued through the Architect. The Architect shall have authority to act on behalf of the Owner to the extent provided in the General Conditions unless otherwise modified in writing.

1.1.13 The Architect shall at all times have access to the Work wherever it is in preparation or progress.

1.1.14 The Architect shall make periodic visits to the site to familiarize himself generally with the progress and quality of the Work and to determine in general if the Work is proceeding in accordance with the Contract Documents. On the basis of his on-site observations as an Architect, he shall endeavor to guard the Owner against defects and deficiencies in the Work of the Contractor. The Architect shall not be required to make exhaustive or continuous on-site inspections to check the quality or quantity of the Work. The Architect shall not be responsible for construction means, methods, techniques, sequences or procedures, or for safety precautions and programs in connection with the Work, and he shall not be responsible for the Contractor's failure to carry out the Work in accordance with the Contract Documents.

1.1.15 Based on such observations at the site and on the Contractor's Applications for Payment, the Architect shall determine the amount owning to the Contractor and shall issue Certificates for Payment in such amounts. The issuance of a Certificate for Payment shall constitute a representation by the Architect to the Owner, based on the Architect's observations at the site as provided in Subparagraph 1.1.14 and on the data comprising the Application for Payment, that the Work has progressed to the point indicated; that to the best of the Architect's knowledge, information and belief, the quality of the Work is in accordance with the Contract Documents (subject to an evaluation of the Work as a functioning whole upon Substantial Completion, to the results of any subsequent tests required by the Contract Documents, to minor deviations from the Contract Documents correctable prior to completion, and to any specific qualifications stated in the Certificate for Payment); and that the Contractor is entitled to payment in the amount certified. By issuing a Certificate for Payment, the Architect shall not be deemed to represent that he has made any examination to ascertain how and for what purpose the Contractor has used the moneys paid on account of the Contract Sum.

1.1.16 The Architect shall be, in the first instance, the interpreter of the requirements of the Contract Documents and the impartial judge of the performance thereunder by both the Owner and Contractor. The Architect shall make decisions on all claims of the Owner or Contractor relating to the execution and progress of the Work and on all other matters or questions related thereto. The Architect's decisions in matters relating to artistic effect shall be final if consistent with the intent of the Contract Documents.

1.1.17 The Architect shall have authority to reject Work which does not conform to the Contract Documents. The Architect shall also have authority to require the Contractor to stop the Work whenever in his reasonable opinion it may be necessary for the proper performance of the Contract. The Architect shall not be liable to the Owner for the consequences of any decision made by him in good faith either to exercise or not to exercise his authority to stop the Work.

1.1.18 The Architect shall review and approve shop drawings, samples, and other submissions of the Contractor only for conformance with the design concept of the Project and for compliance with the information given in the Contract Documents.

1.1.19 The Architect shall prepare Change Orders.

1.1.20 The Architect shall conduct inspections to determine the Dates of Substantial Completion and Final Completion, shall receive written guarantees and related documents assembled by the Contractor, and shall issue a final Certificate for Payment.

1.1.21 The Architect shall not be responsible for the acts or omissions of the Contractor, or any Subcontractors, or any of the Contractor's or Subcontractors' agents or employees, or any other persons performing any of the Work.

1.2 Project Representation Beyond Basic Services

1.2.1 If more extensive representation at the site than is described under Subparagraphs 1.1.10 through 1.1.21 inclusive is required, and if the Owner and Architect agree, the Architect shall provide one or more Full-time Project Representatives to assist the Architect.

1.2.2 Such Full-time Project Representatives shall be selected, employed and directed by the Architect, and the Architect shall be compensated therefor as mutually agreed between the Owner and the Architect as set forth in an exhibit appended to this Agreement.

1.2.3 The duties, responsibilities and limitations of authority of such Full-time Project Representatives shall be set forth in an exhibit appended to this Agreement.

1.2.4 Through the on-site observations by Full-time Project Representatives of the Work in progress, the Architect shall endeavor to provide further protection for the Owner against defects in the Work, but the furnishing of such project representation shall not make the Architect responsible for the Contractor's failure to perform the Work in accordance with the Contract Documents.

1.3 Additional Services

The following services are not covered in Paragraphs 1.1 or 1.2. If any of these Additional Services are authorized by the Owner, they shall be paid for by the Owner as hereinbefore provided.

1.3.1 Providing special analyses of the Owner's needs, and programming the requirements of the Project.

1.3.2   Providing financial feasibility or other special studies.

1.3.3   Providing planning surveys, site evaluations, or comparative studies of prospective sites.

1.3.4   Making measured drawings of existing construction when required for planning additions or alterations thereto.

1.3.5   Revising previously approved Drawings, Specifications or other documents to accomplish changes not initiated by the Architect.

1.3.6   Preparing Change Orders and supporting data where the change in the Basic Fee resulting from the adjusted Contract Sum is not commensurate with the Architect's services required.

1.3.7   Preparing documents for alternate bids requested by the Owner.

1.3.8   Providing Detailed Estimates of Construction Costs.

1.3.9   Providing consultation concerning replacement of any Work damaged by fire or other cause during construction, and furnishing professional services of the type set forth in Paragraph 1.1 as may be required in connection with the replacement of such Work.

1.3.10   Providing professional services made necessary by the default of the Contractor in the performance of the Construction Contract.

1.3.11   Providing Contract Administration and observation of construction after the Contract Time has been exceeded by more than twenty per cent through no fault of the Architect.

1.3.12   Furnishing the Owner a set of reproducible record prints of drawings showing significant changes made during the construction process, based on marked up prints, drawings and other data furnished by the Contractor to the Architect.

1.3.13   Providing services after final payment to the Contractor.

1.3.14   Providing interior design and other services required for or in connection with the selection of furniture and furnishings.

1.3.15   Providing services as an expert witness in connection with any public hearing, arbitration proceeding, or the proceedings of a court of record.

1.3.16   Providing services for planning tenant or rental spaces.

ARTICLE 2

THE OWNER'S RESPONSIBILITIES

2.1   The Owner shall provide full information regarding his requirements for the Project.

2.2   The Owner shall designate, when necessary, a representative authorized to act in his behalf with respect to the Project. The Owner or his representative shall examine documents submitted by the Architect and shall render decisions pertaining thereto promptly, to avoid unreasonable delay in the progress of the Architect's work.

2.3   The Owner shall furnish a certified land survey of the site giving, as applicable, grades and lines of streets, alleys pavements and adjoining property; rights-of-way, restrictions, easements, encroachments, zoning, deed restrictions, boundaries and contours of the site; locations, dimensions and complete data pertaining to existing buildings, other improvements and trees; and full information concerning available service and utility lines both public and private.

2.4 The Owner shall furnish the services of a soils engineer, when such services are deemed necessary by the Architect, including reports, test borings, test pits, soil bearing values and other necessary operations for determining subsoil conditions.

2.5 The Owner shall furnish structural, mechanical, chemical and other laboratory tests, inspections and reports as required by law or the Contract Documents.

2.6 The Owner shall furnish such legal, accounting and insurance counselling services as may be necessary for the Project, and such auditing services as he may require to ascertain how or for what purposes the Contractor has used the moneys paid to him under the Construction Contract.

2.7 The services, information, surveys and reports required by Paragraphs 2.3 through 2.6 inclusive shall be furnished at the Owner's expense, and the Architect shall be entitled to rely upon the accuracy thereof.

2.8 If the Owner observes or otherwise becomes aware of any fault or defect in the Project or non-conformance with the Contract Documents, he shall give prompt written notice thereof to the Architect.

2.9 The Owner shall furnish information required of him as expeditiously as necessary for the orderly progress of the Work.

ARTICLE 3

CONSTRUCTION COST

3.1 Construction Cost to be used as a basis for determining the Architect's Fee for all Work designed or specified by the Architect, including labor, materials, equipment and furnishings, shall be determined as follows, with precedence in the order listed:

3.1.1 For completed construction, the total cost of all such Work;

3.1.2 For Work not constructed, the lowest bona fide bid received from a qualified bidder for any or all of such work; or

3.1.3 For work for which bids are not received, (1) the latest Detailed Cost Estimate, or (2) the Architect's latest Statement of Probable Construction Cost.

3.2 Construction Cost does not include the fees of the Architect and consultants, the cost of the land, rights-of-way, or other costs which are the responsibility of the Owner as provided in Paragraphs 2.3 through 2.6 inclusive.

3.3 Labor furnished by the Owner for the Project shall be included in the Construction Cost at current market rates. Materials and equipment furnished by the Owner shall be included at current market prices, except that used materials and equipment shall be included as if purchased new for the Project.

3.4 Statements of Probable Construction Cost and Detailed Cost Estimates prepared by the Architect represent his best judgment as a design professional familiar with the construction industry. It is recognized, however, that neither the Architect nor the Owner has any control over the cost of labor, materials or equipment, over the contractors' methods of determining bid prices, or over competitive bidding or market conditions. Accordingly, the Architect cannot and does not guarantee that bids will not vary from any Statement of Probable Construction Cost or other cost estimate prepared by him.

3.5   When a fixed limit of Construction Cost is established as a condition of this Agreement, it shall include a bidding contingency of ten per cent unless another amount is agreed upon in writing. When such a fixed limit is established, the Architect shall be permitted to determine what materials, equipment, component systems and types of construction are to be included in the Contract Documents, and to make reasonable adjustments in the scope of the Project to bring it within the fixed limit. The Architect may also include in the Contract Documents alternate bids to adjust the Construction Cost to the fixed limit.

3.5.1   If the lowest bona fide bid, the Detailed Cost Estimate or the Statement of Probable Construction Cost exceeds such fixed limit of Construction Cost (including the bidding contingency) established as a condition of this Agreement, the Owner shall (1) give written approval of an increase in such fixed limit, (2) authorize rebidding the Project within a reasonable time, or (3) cooperate in revising the Project scope and quality as required to reduce the Probable Construction Cost. In the case of (3) the Architect, without additional charge, shall modify the Drawings and Specifications as necessary to bring the Construction Cost within the fixed limit. The providing of this service shall be the limit of the Architect's responsibility in this regard, and having done so, the Architect shall be entitled to his fees in accordance with this Agreement.

ARTICLE 4

DIRECT PERSONNEL EXPENSE

4.1   Direct Personnel Expense of employees engaged on the Project by the Architect includes architects, engineers, designers, job captains, draftsmen, specification writers and typists, in consultation, research and design, in producing Drawings, Specifications and other documents pertaining to the Project, and in services during construction at the site.

4.2   Direct Personnel Expense includes cost of salaries and of mandatory and customary benefits such as statutory employee benefits, insurance, sick leave, holidays and vacations, pensions and similar benefits.

ARTICLE 5

REIMBURSABLE EXPENSES

5.1   Reimbursable Expenses are in addition to the Fees for Basic and Additional Services and include actual expenditures made by the Architect, his employees, or his consultants in the interest of the Project for the following incidental expenses listed in the following Subparagraphs:

5.1.1 Expense of transportation and living when traveling in connection with the Project and for long distance calls and telegrams.

5.1.2 Expense of reproductions, postage and handling of Drawings and Specifications, excluding copies for Architect's office use and duplicate sets at each phase for the Owner's review and approval; and fees paid for securing approval of authorities having jurisdiction over the Project.

5.1.3 If authorized in advance by the Owner, the expense of overtime work requiring higher than regular rates; perspectives or models for the Owner's use; and fees of special consultants for other than the normal structural, mechanical and electrical engineering services.

ARTICLE 6

PAYMENTS TO THE ARCHITECT

6.1 Payments on account of the Architect's Basic Services shall be made as follows:

6.1.1 An initial payment of five per cent of the Basic Fee calculated upon an agreed estimated cost of the Project, payable upon execution of this Agreement, is the minimum payment under this Agreement.

6.1.2 Subsequent payments shall be made monthly in proportion to services performed to increase the compensation for Basic Services to the following percentages of the Basic Fee at the completion of each phase of the Work:

Schematic Design Phase	15%
Design Development Phase	35%
Construction Documents Phase	75%
Bidding or Negotiation Phase	80%
Construction Phase	100%

6.2 Payments for Additional Services of the Architect as defined in Paragraph 1.3, and for Reimbursable Expenses as defined in Article 5, shall be made monthly upon presentation of the Architect's statement of services rendered.

6.3 No deductions shall be made from the Architect's compensation on account of penalty, liquidated damages, or other sums withheld from payments to contractors.

6.4 If the Project is suspended for more than three months or abandoned in whole or in part, the Architect shall be paid his compensation for services performed prior to receipt of written notice from the Owner of such suspension or abandonment, together with Reimbursable Expenses then due and all terminal expenses resulting from such suspension or abandonment.

ARTICLE 7

ARCHITECT'S ACCOUNTING RECORDS

Records of the Architect's Direct Personnel, Consultant and Reimbursable Expenses pertaining to the Project, and records of accounts between the Owner and the Contractor, shall be kept on a generally recognized accounting basis and shall be available to the Owner or his authorized representative at mutually convenient times.

ARTICLE 8

TERMINATION OF AGREEMENT

This Agreement may be terminated by either party upon seven days' written notice should the other party fail substantially to perform in accordance with its terms through no fault of the other. In the event of termination due to the fault of others than the Architect, the Architect shall be paid his compensation for services performed to termination date, including Reimbursable Expenses then due and all terminal expenses.

ARTICLE 9

OWNERSHIP OF DOCUMENTS

Drawings and Specifications as instruments of service are and shall remain the property of the Architect whether the Project for which they are made is executed or not. They are not to be used by the Owner on other projects or extensions to this Project except by agreement in writing and with appropriate compensation to the Architect.

ARTICLE 10

SUCCESSORS AND ASSIGNS

The Owner and the Architect each binds himself, his partners, successors, assigns and legal representatives to the other party to this Agreement and to the partners, successors, assigns and legal representatives of such other party with respect to all covenants of this Agreement. Neither the Owner nor the Architect shall assign, sublet or transfer his interest in this Agreement without the written consent of the other.

ARTICLE 11

ARBITRATION

11.1 All claims, disputes and other matters in question arising out of, or relating to, this Agreement or the breach thereof shall be decided by arbitration in accordance with the Construction Industry Arbitration Rules of the American Arbitration Association then obtaining. This agreement so to arbitrate shall be specifically enforceable under the prevailing arbitration law.

11.2 Notice of the demand for arbitration shall be filed in writing with the other party to this Agreement and with the American Arbitration Association. The demand shall be made within a reasonable time after the claim, dispute or other matter in question has arisen. In no event shall the demand for arbitration be made after institution of legal or equitable proceedings based on such claim, dispute or other matter in question would be barred by the applicable statute of limitations.

11.3 The award rendered by the arbitrators shall be final, and judgment may be entered upon it in any court having jurisdiction thereof.

ARTICLE 12

EXTENT OF AGREEMENT

This Agreement represents the entire and integrated agreement between the Owner and the Architect and supersedes all prior negotiations, representations or agreements, either written or oral. This Agreement may be amended only by written instrument signed by both Owner and Architect.

ARTICLE 13

APPLICABLE LAW

Unless otherwise specified, this Agreement shall be governed by the law of the principal place of business of the Architect.

This Agreement executed the day and year first written above.

OWNER _____   ARCHITECT _____

Architect's Registration No. _____

AWARD/CONTRACT

STANDARD FORM 26, JULY 1966
GENERAL SERVICES ADMINISTRATION
FED. PROC. REG. (41CFR) 1-16.101

PAGE 1 OF

1. CONTRACT (Proc. Inst. Ident.) NO.

2. EFFECTIVE DATE

3. REQUISITION/PURCHASE REQUEST/PROJECT NO.

4. CERTIFIED FOR NATIONAL DEFENSE UNDER BSDA REG. 2 AND/OR DMS REG. 1. RATING:

5. ISSUED BY   CODE

6. ADMINISTERED BY (If other than block 5)   CODE

7. DELIVERY
- [ ] FOB DESTINATION
- [ ] OTHER (See below)

8. CONTRACTOR NAME AND ADDRESS   CODE

(Street, city, county, State, and ZIP code)

FACILITY CODE

9. DISCOUNT FOR PROMPT PAYMENT

10. SUBMIT INVOICES (4 copies unless otherwise specified) TO ADDRESS SHOWN IN BLOCK

11. SHIP TO/MARK FOR   CODE

12. PAYMENT WILL BE MADE BY   CODE

13. THIS PROCUREMENT WAS [ ] ADVERTISED, [ ] NEGOTIATED, PURSUANT TO:
- [ ] 10 U.S.C. 2304 (a)( )
- [ ] 41 U.S.C. 252 (c)( )

14	Appropriation and Subhead	Obj. Cl.	Bureau Cont. No.	Sub-Allot.	Authorization Acct'g Act'y	Trans. Type	Property Acct'g Act'y	Country

15. ITEM NO.	16. SUPPLIES/SERVICES	17. QUANTITY	18. UNIT	19. UNIT PRICE		20. AMOUNT
				Cost Code	Amount	

21.   TOTAL AMOUNT OF CONTRACT $

CONTRACTING OFFICER WILL COMPLETE BLOCK 22 OR 26 AS APPLICABLE

22. ☐ CONTRACTOR'S NEGOTIATED AGREEMENT *(Contractor is required to sign this document and return _____ copies to issuing office.)* Contractor agrees to furnish and deliver all items or perform all the services set forth or otherwise identified above and on any continuation sheets for the consideration stated herein. The rights and obligations of the parties to this contract shall be subject to and governed by the following documents: (a) this award/contract, (b) the solicitation, if any, and (c) such provisions, representations, certifications, and specifications, as are attached or incorporated by reference herein. *(Attachments are listed herein.)*

23. NAME OF CONTRACTOR

BY

_____
(Signature of person authorized to sign)

24. NAME AND TITLE OF SIGNER *(Type or print).*

25. DATE SIGNED

26. ☐ AWARD *(Contractor is not required to sign this document.)* Your offer on Solicitation Number _____, including the additions or changes made by you which additions or changes are set forth in full above, is hereby accepted as to the items listed above and on any continuation sheets. This award consummates the contract which consists of the following documents: (a) the Government's solicitation and your offer, and (b) this award/contract. No further contractual document is necessary.

27. UNITED STATES OF AMERICA

BY

_____
(Signature of Contracting Officer)

28. NAME OF CONTRACTING OFFICER *(Type or print)*

29. DATE SIGNED

NAVFAC 4-4330/11(10-67)
Supersedes NavDocks 424 (Rev. 4−65)

SCHEDULE

It has been determined that the execution of this contract is advantageous to the national defense and that the existing facilities of the Naval Establishment are inadequate.

1. DESCRIPTION OF THE PROJECT. The Architect-Engineer shall perform all the services required under this contract for the project generally described as follows: (hereinafter refered to as the "Project")

_____

_____

having an estimated construction cost of $_____, and more specifically described in APPENDIX "A" which is attached hereto and made a part hereof.

2. STATEMENT OF ARCHITECT-ENGINEER SERVICES. The Architect-Engineer shall prepare and furnish to the Government, complete and ready for use, all necessary studies, preliminary sketches, estimates, working records and other drawings (including large scale details as required), and specifications; shall check shop drawings furnished by the construction contractor; shall furnish consultation and advice as requested by the Government during the construction (but not including the supervision of the construction work); and shall furnish all other architectural and engineering services, including without limitations those specified hereinafter and required in connection with the accomplishment of Naval public works and/or utilities projects.

3. PERIOD OF SERVICE. The Architect-Engineer shall complete all work and services within _____ months after receipt of notice to proceed.

4. DESIGN WITHIN COST LIMITATION. The Architect-Engineer agrees to design a facility which can be constructed (under normal Bureau of Yards and Docks procedures) within the estimated net construction cost, with each item of the facility to be designed so as to be constructed within applicable statutory cost limitations indicated below.

If, after receipt of competitive bids, it is found that a construction contract cannot be awarded within the aforementioned estimated construction cost and within any applicable statutory cost limitations, the Architect-Engineer will, as a part of this contract, and at no additional expense to the Government, perform such redesign, re-estimating and other services as may, in the opinion of the Officer in Charge, be required to produce a usable facility within the estimated construction cost and/or within the statutory cost limitations. In connection with the foregoing, the Architect-Engineer shall be obligated to perform such additional services at no increase in contract price only when the Navy has received competitive bids within six (6) months from the date of final approval of the plans and specifications.

STATUTORY COST LIMITS

_____

5. ARCHITECTURAL DESIGNS AND DATA—GOVERNMENT RIGHTS (UN-LIMITED) (APR.1966). The Government shall have unlimited rights, for the benefit of the Government, in all drawings, designs, specifications, architectural designs of buildings and structures, notes and other architect-engineer work produced in the performance of this contract, or in contemplation thereof, and all as-built drawings produced after completion of the work, including the right to use same on any other Government work without additional cost to the Government; and with respect thereto the Architect-Engineer agrees to and does hereby grant to the Government a royalty-free license to all such data which he may cover by copyright and to all architectural designs as to which he may assert any rights or establish any claim under the design patent or copyright laws. The Architect-Engineer for a period of three (3) years after completion of the project agrees to furnish and to provide access to the originals or copies of all such materials on the request of the Contracting Officer.

6. ADDITIONAL STATEMENT OF ARCHITECT-ENGINEER SERVICES. It is agreed, without limiting the generality of the foregoing, that (a) The Architect-Engineer shall, if necessary, visit the site and shall hold such conferences with representatives of the Government and take such other action as may be necessary to obtain the data upon which to develop the design and preliminary sketches showing the contemplated project, (b) The preliminary sketches shall include plans, elevations and sections developed in such detail and with such descriptive specifications as will clearly indicate the scope of the work, and make possible a reasonable estimate of the cost, (c) Preliminary sketches together with an estimate of the cost of the project shown on the sketches shall be submitted for the approval of the Officer in Charge, (d) The Architect-Engineer shall change the preliminary sketches for the extent necessary to meet the requirements of the Government, and after approval by the Officer in Charge the Architect-Engineer shall furnish necessary prints of the approved preliminary sketches to the Officer in Charge, (e) After the preliminary sketches and estimates have been approved, the Architect-Engineer shall proceed with the preparation of complete working drawings and specifications as required by the Officer in Charge in connection with the construction of the said project. Working drawings, specifications and estimates shall be delivered to the Officer in Charge in such sequence and at such times as required by the Government and as will insure that the construction work can be initiated promptly, procurement of materials made without delay, and continuous prosecution of the work promoted. Working drawings and specifications shall be revised as necessary and as required by the Officer in Charge. After working drawings and specifications have been approved by the Officer in Charge, the Architect-Engineer shall furnish such number of sets of prints of the approved working drawings and such number of sets of the approved specifications, as may be required by the Officer in Charge (f) Upon approval of final plans the Architect-Engineer shall deliver to the Government one set of tracings, in such medium and on such materials, as may be required by the Officer in Charge, suitable for blueprinting, showing complete approved construction requirements (not of "as-built" construction unless otherwise stipulated), provided, however, that should this contract be terminated by the Government, the Architect-Engineer shall deliver to the Government one set of such tracings as may be required by the Officer in Charge. Such tracings as are delivered shall be signed by the Officer in Charge as an indication of approval thereof, and shall become and remain the property of the Government. (g) The Architect-Engineer shall perform all necessary architectural-engineering services of every king required in connection with the studies, designs and the preparation of drawings and specifications, but said services shall not, unless otherwise stipulated, include borings, test piles and pits, or supervision of construction work executed from the drawings and specifications, provided, however, that the Architect-Engineer shall furnish upon request and without additional compensation, such

amplifications and explanations and attend such conferences as may, in the opinion of the Officer-in-Charge, be necessary to clarify the intent of the drawings and specifications and shall afford the benefit of his advice on questions that may arise in connection with the construction of the project. (h) The Architect-Engineer shall without additional fee correct or revise the drawings, specifications, or other materials furnished under this contract, if the Officer in Charge finds that such revision is necessary to correct errors or deficiencies for which the Architect-Engineer is responsible.

7. DEFINITIONS (JUNE 1964). (a) The term "head of the agency" or "Secretary" as used herein means the Secretary, the Under Secretary, and Assistant Secretary, or any other head or assistant head of the executive or military department or other Federal agency; and the term "his duly authorized representative" means any person or persons or board (other than the Contracting Officer) authorized to act for the head of the agency or the Secretary.

(b) The term "Contracting Officer" as used herein means the person executing this contract on behalf of the Government and includes a duly appointed successor or authorized representative. (See "Government Representatives" clause).

(c) Except as otherwise provided in this contract, the term "subcontracts" includes purchase orders under this contract.

8. GOVERNMENT REPRESENTATIVES. The work will be under the general direction of the Contracting Officer, The Commander of Naval Facilities Engineering Command, who shall designate an officer of the Civil Engineer Corps, United States Navy, or other officer or representative of the Government, as Officer in Charge, hereinafter referred to as the "OIC" who, except in connection with the Disputes Clause shall be the authorized representative of the Contracting Officer and under the direction of the Contracting Officer have complete charge of the work, and shall exercise full supervision and general direction of the work, so far as it affects the interest of the Government. For the purposes of the "Disputes" clause the Contracting Officer shall mean the Commander, Naval Facilities Engineering Command, the Acting Commander, their successors, or their representatives specially designated for this purpose.

9. CONTRACTING OFFICER'S DECISIONS (JAN. 1965). The extent and character of the work to be done by the Architect-Engineer shall be subject to the general supervision, direction, control and approval of the Contracting Officer.

10. TECHNICAL ADEQUACY. Approval by the Officer in Charge of drawings, designs, specifications and other incidental architectural engineering work or materials furnished hereunder shall not in any way relieve the Architect-Engineer of responsibility for the technical adequacy of the work.

11. COMPENSATION. The Architect-Engineer shall be paid the lump sum of $ as full compensation for all services, labor and material required hereby, and for all expenditures which may be made and expenses incurred except as are otherwise expressly provided herein.

12. METHOD OF PAYMENT (JAN. 1965). (a) Estimates shall be made monthly of the amount and value of the work and services performed by the Architect-Engineer under this contract, such estimates to be prepared by the Architect-Engineer and accompanied by such supporting data as may be required by the Contracting Officer.

(b) Upon approval of such estimate by the Contracting Officer payment upon properly certified vouchers shall be made to the Architect-Engineer as soon as practicable of

90% of the amount as determined above, less all previous payments; provided, however, that if the Contracting Officer determines that the work is substantially complete and that the amount of retained percentages is in excess of the amount considered by him to be adequate for the protection of the Government, he may at his discretion release to the Architect-Engineer such excess amount.

(c) Upon satisfactory completion by the Architect-Engineer and acceptance by the Contracting Officer of the work done by the Architect-Engineer in accordance with the foregoing "Statement of Architect-Engineer Services," the Architect-Engineer will be paid the unpaid balance of any money due for work under said statement, including retained percentages relating to this portion of the work.

(d) Upon satisfactory completion of the construction work and its final acceptance, the Architect-Engineer shall be paid the unpaid balance of any money due hereunder. Prior to such final payment under this contract, or prior settlement upon termination of the contract, and as a condition precedent thereto, the Architect-Engineer shall execute and deliver to the Contracting Officer a release of all claims against the Government arising under or by virtue of this contract, other than such claims, if any, as may be specifically excepted by the Architect-Engineer from the operation of the release in stated amounts to be set forth therein.

13. CHANGES (JAN. 1965). The Contracting Officer may at any time, by written order, and without notice to the sureties, make changes within the general scope of the contract in the work and service to be performed. If any such changes cause an increase or decrease in the Architect-Engineer's cost of, or the time required for, performance of this contract, an equitable adjustment shall be made and the contract shall be modified in writing accordingly. Any claim by the Architect-Engineer for adjustment under this clause must be asserted in writing within 30 days from the date of receipt by the Architect-Engineer of the notification of change unless the Contracting Officer grants a further period of time before the date of final payment under the contract. If the parties fail to agree upon the adjustment to be made, the dispute shall be determined as provided in the "Disputes" clause of this contract; but nothing provided in this clause shall excuse the Architect-Engineer from proceeding with the prosecution of the work as changed. Except as otherwise provided in this contract, no charge for any extra work or material will be allowed.

14. FEDERAL, STATE, AND LOCAL TAXES (JUL. 1960). (a) As used throughout this clause, the term "contract date" means the date of this contract. As to additional supplies or services procured by modification of this contract, the term "contract date" means the date of such modification.

(b) Except as may be otherwise provided in this contract, the contract price includes, to the extent allocable to this contract, all Federal, State, and local taxes which, on the contract date:

(i) by Constitution, statute, or ordinance, are applicable to this contract, or to the transactions covered by this contract, or to property or interests in property; or

(ii) pursuant to written ruling or regulation, the authority charged with administering any such tax is assessing or applying to, and is not granting or honoring an exemption for, a contractor under this kind of contract, or the transactions covered by this contract, or property or interests in property.

(c) Except as may be otherwise provided in this contract, duties in effect on the contract date are included in the contract price, to the extent allocable to this contract.

(d) (1) If the Architect-Engineer is required to pay or bear the burden—

(i) of any tax or duty which either was not to be included in the contract price pursuant to the requirements of paragraphs (b) and (c), or of a tax or duty specifically excluded from the contract price by a provision of this contract; or

(ii) of an increase in rate of any tax or duty, whether or not such tax or duty was excluded from the contract price; or

(iii) of any interest or penalty on any tax or duty referred to in (i) or (ii) above; the contract price shall be increased by the amount of such tax, duty, interest, or penalty allocable to this contract; *provided,* that the Architect-Engineer warrants in writing that no amount of such tax, duty, or rate increase was included in the contract price as a contingency reserve or otherwise; and *provided further,* that liability for such tax, duty, rate increase, interest, or penalty was not incurred through the fault or negligence of the Architect-Engineer or his failure to follow instructions of the Contracting Officer.

(2) If the Architect-Engineer is not required to pay or bear the burden, or obtains a refund or drawback, in whole or in part, of any tax, duty, interest, or penalty which:

(i) was to be included in the contract price pursuant to the requirements of paragraphs (b) and (c);

(ii) was included in the contract price; or

(iii) was the basis of an increase in the contract price; the contract price shall be decreased by the amount of such relief, refund, or drawback allocable to this contract, or the allocable amount of such relief, refund, or drawback shall be paid to the Government, as directed by the Contracting Officer. The contract price also shall be similarly decreased if the Contractor, through his fault or negligence or his failure to follow instructions of the Contracting Officer, is required to pay or bear the burden, or does not obtain a refund or drawback of any such tax, duty, interest, or penalty. Interest paid or credited to the Architect-Engineer incident to a refund of taxes shall inure to the benefit of the Government to the extent that such interest was earned after the Architect-Engineer was paid or reimbursed by the Government for such taxes.

(3) Invoices or vouchers covering any adjustment of the contract price pursuant to this paragraph (d) shall set forth the amount thereof as a separate item and shall identify the particular tax or duty involved.

(4) This paragraph (d) shall not be applicable to social security taxes; income and franchise taxes, other than those levied on or measured by (i) sales or receipts from sales, or (ii) the Architect-Engineer's possession of, interest in, or use of property, title to which is in the Government; excess profits taxes; capital stock taxes; unemployment compensation taxes; or property taxes, other than such property taxes, allocable to this contract, as are assessed either on completed supplies covered by this contract, or on the Architect-Engineer's possession of, interest in, or use of property, title to which is in the Government.

(5) No adjustment of less than $100 is required to be made in the contract price pursuant to this paragraph (d).

(e) Unless there does not exist any reasonable basis to sustain an exemption, the Government upon request of the Architect-Engineer, without further liability, agrees, except as otherwise provided in this contract, to furnish evidence appropriate to establish exemption from any tax which the Architect-Engineer warrants in writing was excluded from the contract price. In addition, the Contracting Officer may furnish evidence appropriate to establish exemption from any tax that may, pursuant to this clause, give rise to either an increase or decrease in the contract price. Except as otherwise provided in this contract, evidence appropriate to establish exemption from duties will be furnished only at the discretion of the Contracting Officer.

(f) (1) The Architect-Engineer shall promptly notify the Contracting Officer on all matters pertaining to Federal, State, and local taxes, and duties, that reasonably may be expected to result in either an increase or decrease in the contract price.

(2) Whenever an increase or decrease in the contract price may be required under this clause, the Architect-Engineer shall take action as directed by the Contracting Officer, and the contract price shall be equitably adjusted to cover the costs of such action, including any interest, penalty, and reasonable attorneys' fees.

15. RESPONSIBILITY OF THE ARCHITECT-ENGINEER. (a) The Architect-Engineer shall be responsible for the professional and technical accuracy and the coordination of all designs, drawings, specifications and other work or materials furnished by the Architect-Engineer under this contract. The Architect-Engineer shall without additional cost or fee to the Government correct or revise any errors or deficiencies in his performance.

(b) Neither the Government's review, approval or acceptance of, nor payment for, any of the services required under this contract shall be construed to operate as a waiver of any rights under this contract or of any cause of action arising out of the performance of this contract, and the Architect-Engineer shall be and remain liable to the Government for all costs of any kind which were incurred by the Government as a result of the Architect-Engineer's negligent performance of and of the services furnished under this contract.

(c) The rights and remedies of the Government provided for under this contract are in addition to any other rights and remedies provided by law; the Government may assert its right of recovery by any appropriate means including, but not limited to, set-off, suit, withholding, recoupment, or counterclaim, either during or after performance of this contract.

16. MILITARY SECURITY REQUIREMENTS (APR.1966). (a) The provisions of this clause shall apply to the extent that this contract involves access to information classified "Confidential" including "Confidential–Modified Handling Authorized" or higher.

(b) The Government shall notify the Architect-Engineer of the security classification of this contract and the elements thereof, and of any subsequent revisions in such security classification, by the use of a Security Requirements Check List (DD Form 254), or other written notification.

(c) To the extent the Government has indicated as of the date of this contract or thereafter indicates security classification under this contract as provided in paragraph (b) above, the Architect-Engineer shall safeguard all classified elements of this contract and shall provide and maintain a system of security controls within its own organization in accordance with the requirements of:

(i) the Security Agreement (DD Form 441), including the Department of Defense Industrial Security Manual for Safeguarding Classified Information as in effect on the date of this contract, and any modification to the Security Agreement for the purpose of adapting the Manual to the Architect-Engineer's business; and

(ii) any amendments to said Manual made after the date of this contract, notice of which has been furnished to the Architect-Engineer by the Security Office of the Military Department having security cognizance over the facility.

(d) Representatives of the Military Department having security cognizance over the facility and representatives of the contracting Military Department shall have the right to inspect at reasonable intervals the procedures, methods, and facilities utilized by the Architect-Engineer in complying with the security requirements under this contract. Should the Government, through these representatives, determine that the Architect-Engineer is not complying with the Security requirements of this contract the Architect-Engineer shall be informed in writing by the Security Office of the cognizant Military Department of the proper action to be taken in order to effect compliance with such requirements.

(e) If, subsequent to the date of this contract, the security classifications or security requirements under this contract are changed by the Government as provided in this clause and the security costs or time required for delivery under this contract are thereby increased or decreased, the contract price, delivery schedule, or both and any other provision of the contract that may be affected shall be subject to an equitable adjustment by reason of such increased or decreased costs. Any equitable adjustment shall be accomplished in the same manner as if such changes were directed under the "Changes" clause in this contract.

(f) The Architect-Engineer agrees to insert, in all subcontracts hereunder which involve access to classified information, provisions which shall conform substantially to the language of this clause, including this paragraph (f) but excluding the last sentence of paragraph (e) of this clause.

(g) The Architect-Engineer also agrees that it shall determine that any subcontractor proposed by it for the furnishing of supplies and services which involve access to classified information in the Architect-Engineer's custody has been granted an appropriate facility security clearance which is still in effect, prior to being accorded access to such classified information.

17. TERMINATION FOR DEFAULT (JAN. 1965). (a) The performance of work under this contract may be terminated by the Government in accordance with this clause in whole, or from time to time in part, whenever the Architect-Engineer shall default in performance of this contract in accordance with its terms (including in the term "default" any such failure by the Architect-Engineer to make progress in the prosecution of the work hereunder as endangers such performance), and shall fail to cure such default within a period of ten days (or such longer periods as the Contracting Officer may allow) after receipt by the Architect-Engineer of a notice specifying the default.

(b) If the contract is so terminated, the Government may take over the work and services and prosecute the same to completion by contract or otherwise, and the Architect-Engineer shall be liable to the Government for any excess cost occasioned to the Government thereby.

(c) The contract may not be so terminated if the failure to perform arises from unforseeable causes beyond the control and without the fault or negligence of the Architect-Engineer. Such causes may include, but are not restricted to acts of God, acts of the public enemy, acts of the Government in either its sovereign or contractural capacity, fires, floods, epidemics, quarantine restrictions, strikes and unusually severe weather; but in every case the failure to perform must be beyond the control and without the fault or negligence of the Architect-Engineer; and the Architect-Engineer, within ten days from the beginning of any such delay (unless the Contracting Officer grants a further period of time before the date of final payment under the contract) notifies the Contracting Officer in writing of the causes of delay. The Contracting Office shall ascertain the facts and the extent of delay and extend the time for completing the work when, in his judgment, the findings of fact justify such an extension, and his finding of fact shall be final and conclusive on the parties, subject only to appeal as provided in the clause of this contract entitled "Disputes."

(d) If, after Notice of Termination of the contract under the provisions of this clause, it is determined for any reason that the Architect-Engineer was not in default under the provisions of this clause or that the default was excusable under the provisions of this clause, the rights and obligations of the parties shall be the same as if the Notice of Termination had been issued pursuant to the clause of this contract entitled "Termination for the Convenience of the Government."

(e) The rights and remedies of the Government provided in this clause are in addition to any other rights and remedies provided by law or under this contract.

18. DISPUTES (JUNE 1964). (a) Except as otherwise provided in this contract, any dispute concerning a question of fact arising under this contract which is not disposed of by agreement shall be decided by the Contracting Officer, who shall reduce his decision to writing and mail or otherwise furnish a copy thereof to the Architect-Engineer. The decision of the Contracting Officer shall be final and conclusive unless, within 30 days from the date of receipt of such copy, the Architect-Engineer mails or otherwise furnishes to the Contracting Officer a written appeal addressed to the head of the agency involved. The decision of the head of the agency or his duly authorized representative for the determination of such appeals shall be final and conclusive. This provision shall not be pleaded in any suit involving a question of fact arising under this contract as limiting judicial review of any such decision to cases where fraud by such official or his representative or board is alleged: Provided, however, that any such decision shall be final and conclusive unless the same is fraudulent or capricious or arbitrary or so grossly erroneous as necessarily to imply bad faith or is not supported by substantial evidence. In connection with any appeal proceeding under this clause, the Architect-Engineer shall be afforded an opportunity to be heard and to offer evidence in support of his appeal. Pending final decision of a dispute hereunder, the Architect-Engineer shall proceed diligently with the performance of the contract and in accordance with the Contracting Officer's decision.

(b) This Disputes clause does not preclude consideration of questions of law in connection with decisions provided for in paragraph (a) above. Nothing in this contract, however, shall be construed as making final the decision of any administrative official, representative, or board on a question of law.

19. OFFICIALS NOT TO BENEFIT (JUNE 1964). No member of or delegate to Congress, or resident commissioner, shall be admitted to any share or part of this contract, or to any benefit that may arise therefrom; but this provision shall not be construed to extend to this contract if made with a corporation for its general benefit.

20. COVENANT AGAINST CONTINGENT FEES (JAN. 1958). The Architect-Engineer warrants that no person or selling agency has been employed or retained to solicit or secure this contract upon an agreement or understanding for a commission, percentage, brokerage, or contingent fee, excepting bona fide employees or bona fide established commercial or selling agencies maintained for the Architect-Engineer for the purpose of securing business. For breach or violation of this warranty the Government shall have the right to annul this contract without liability or in its discretion to deduct from the contract price or consideration the full amount of such commission, percentage, brokerage, or contingent fee.

21. ASSIGNMENT OF CLAIMS (JUNE 1964). (a) Pursuant to the provisions of the Assignment of Claims Act of 1940 as amended (31 U.S. Code 203, 41 U.S. Code 15), if this

contract provides for payments aggregating $1,000 or more, claims for moneys due or to become due the Architect-Engineer from the Government under this contract may be assigned to a bank, trust company or other financing institution, including any Federal lending agency, and may thereafter be further assigned and reassigned to any such institution. Any such assignment or reassignment shall cover all amounts payable under this contract and not already paid, and shall not be made to more than one party, except that any such assignment or reassignment may be made to one party as agent or trustee for two or more parties participating in such financing. Unless otherwise provided in this contract, payments to an assignee of any moneys due or to become due under this contract shall not, to the extent provided in said Act as amended, be subject to reduction or set-off.

(b) In no event shall copies of this contract or of any plans, specifications, or other similar documents relating to work under this contract, if marked "Top Secret," "Secret," or "Confidential" be furnished to any assignee of any claim arising under this contract or to any other person not entitled to receive the same. However a copy of any part or all of this contract so marked may be furnished, or any information contained therein may be disclosed, to such assignee upon the prior written authorization of the Contracting Officer.

22. CONTRACT WORK HOURS STANDARDS ACT–OVERTIME COMPENSATION (40 U.S.C. 327-330) (APR. 1965). *(This clause is only applicable to contracts exceeding $2,500 and is required by 40 USC 327-330).* (a) The Architect-Engineer shall not require or permit any laborer or mechanic in any workweek in which he is employed on any work under this contract to work in excess of eight (8) hours in any calendar day or in excess of forty (40) hours in such workweek on work subject to the provisions of the Contract Work Hours Standards Act unless such laborer or mechanic receives compensation at a rate not less than one and one-half times his basic rate of pay for all such hours worked in excess of eight (8) hours in any calendar day or in excess of forty (40) hours in such workweeks, whichever is the greater number of overtime hours. The "basic rate of pay," as used in this clause, shall be the amount paid per hour, exclusive of the Architect-Engineer contribution or cost for fringe benefits and any cash payment made in lieu of providing fringe benefits, or the basic hourly rate contained in the wage determination, whichever is greater.

(b) In the event of any violation of the provisions of paragraph (a), the Architect-Engineer shall be liable to any affected employee for any amounts due, and to the United States for liquidated damages. Such liquidated damages shall be computed with respect to each individual laborer or mechanic employed in violation of the provisions of paragraph (a) in the sum of $10 for each calendar day in which such employee was required or permitted to be employed on such work in excess of eight (8) hours or in excess of the standard workweek of forty (40) hours without payment of the overtime wages required by paragraph (a).

23. CONVICT LABOR (MAR 1949). In connection with the performance of work under this contract, the Architect-Engineer agrees not to employ any person undergoing sentence of imprisonment at hard labor.

24. EQUAL OPPORTUNITY (APRIL 1964). (The following clause is applicable unless this contract is exempt under the rules and regulations of the President's Committee on Equal Employment Opportunity (41 C.F.R. Chapter 60). Exemptions include contracts and subcontracts (i) not exceeding $10,000, (ii) not exceeding $100,000 for standard commercial supplies or raw materials, and (iii) under which work is performed outside the United States and no recruitment of workers within the United States is involved).

During the performance of this contract, the Architect-Engineer agrees as follows:

(a) The Architect-Engineer will not discriminate against any employee or applicant for employment because of race, creed, color, or national origin. The Architect-Engineer will take affirmative action to ensure that applicants are employed, and that employees are treated during employment, without regard to their race, creed, color, or national origin. Such action shall include, but not be limited to, the following: employment, upgrading, demotion or transfer; recruitment or recruitment advertising; layoff or termination; rates of pay or other forms of compensation; and selection for training, including apprenticeship. The Architect-Engineer agrees to post in conspicuous places, available to employees and applicants for employment, notices to be provided by the Contracting Officer setting forth the provisions of this nondiscrimination clause.

(b) The Architect-Engineer will, in all solicitations or advertisements for employees placed by or on behalf of the Architect-Engineer, state that all qualified applicants will receive consideration for employment without regard to race, creed, color, or national origin.

(c) The Architect-Engineer will send to each labor union or representative of workers with which he has a collective bargaining agreement or other contract or understanding, a notice, to be provided by the Agency Contracting Officer, advising the said labor union of workers' representative of the Architect-Engineer's commitments under this nondiscrimination clause, and shall post copies of the notice in conspicuous places available to employees and applicants for employment.

(d) The Architect-Engineer will comply with all provisions of Executive Order No. 10925 of March 6, 1961, as amended, and of the rules, regulations, and relevant orders of the President's Committee on Equal Employment Opportunity created thereby.

(e) The Architect-Engineer will furnish all information and reports required by Executive Order No. 10925 of March 6, 1961, as amended, and by the rules, regulations, and orders of the said Committee, or pursuant thereto, and will permit access to his books, records, and accounts by the contracting agency and the Committee for purposes of investigation to ascertain compliance with such rules, regulations, and orders.

(f) In the event of the Architect-Engineer's noncompliance with the nondiscrimination clause of this contract or with any of the said rules, regulations, or orders, this contract may be cancelled, terminated, or suspended in whole or in part and the Architect-Engineer may be declared ineligible for further Government contracts in accordance with procedures authorized in Executive Order No. 10925 of March 6, 1961, as amended, and such other sanctions may be imposed and remedies invoked as provided in the said Executive Order or by rule regulation, or order of the President's Committee on Equal Employment Opportunity, or as otherwise provided by law.

(g) The Architect-Engineer will include the provisions of paragraphs (a) through (g) in every subcontract or purchase order unless exempted by rules, regulations, or orders of the President's Committee on Equal Employment Opportunity issued pursuant to Section 303 of Executive Order No. 10925 of March 6, 1961, as amended, so that such provisions will be binding upon each subcontractor or vendor.* The Architect-Engineer will take such action with respect to any subcontract or purchase order as the contracting

*Unless otherwise provided, the "Equal Opportunity" clause is not required to be inserted in subcontracts below the second tier, except for subcontracts involving the performance of "construction work" as the "site of construction" (as those terms are defined in the Committee's rules and regulations) in which case the clause must be inserted in all such contracts. Subcontracts may incorporate by reference the "Equal Opportunity" clause.

agency may direct as a means of enforcing such provisions, including sanctions for non-compliance: *Provided, however,* that in the event the Architect-Engineer becomes involved in, or is threatened with, litigation with a subcontractor or vendor as a result of such direction by the contracting agency, the Architect-Engineer may request the United States to enter into such litigation to protect the interest of the United States.

25. RENEGOTIATION (OCT. 1959). (a) To the extent required by law, this contract is subject to the Renegotiation Act of 1951 (50 U.S.C. App. 1211, et seq.), as amended, and to any subsequent act of Congress providing for the renegotiation of Contracts. Nothing contained in this clause shall impose any renegotiation obligation with respect to this contract or any subcontract hereunder which is not imposed by an act of Congress heretofore or hereafter enacted. Subject to the foregoing this contract shall be deemed to contain all the provisions required by Section 104 of the Renegotiation Act of 1951, and by any such other act without subsequent contract amendment specifically incorporating such provisions.

(b) The Architect-Engineer agrees to insert the provisions of this clause, including this paragraph (b), in all subcontracts, as that term is defined in Section 103g of the Renegotiation Act of 1951, as amended.

26. EXAMINATION OF RECORDS (FEB. 1962). (a) The Architect-Engineer agrees that the Comptroller General of the United States or any of his duly authorized representatives shall, until the expiration of three years after final payment under this contract, have access to and the right to examine any directly pertinent books, documents, papers, and records of the Architect-Engineer involving transactions related to this contract.

(b) The Architect-Engineer further agrees to include in all his subcontracts hereunder a provision to the effect that the subcontractor agrees that the Comptroller General of the United States or any of his duly authorized representative shall, until the expiration of three years after final payment under the subcontract, have access to and the right to examine any directly pertinent books, documents, papers, and records of such subcontractor involving transactions related to the subcontract. The term "subcontract" as used in this clause excludes (i) purchase orders not exceeding $2,500 and (ii) subcontracts or purchase orders for public utility services at rates established for uniform applicability to the general public.

27. GRATUITIES (MAR. 1952). (a) The Government may, by written notice to the Architect-Engineer, terminate the right of the Architect-Engineer to proceed under this contract if it is found, after notice and hearing, by the Secretary or his duly authorized representative, that gratuities (in the form of entertainment, gifts, or otherwise) were offered or given by the Architect-Engineer, or any agent or representative of the Architect-Engineer, to any officer or employee of the Government with a view toward securing a contract or securing favorable treatment with respect to the awarding or amending, or the making of any determinations with respect to the performing, of such contract; provided, that the existence of the facts upon which the Secretary or his duly authorized representatives makes such finding shall be in issue and may be reviewed in any competent court.

(b) In the event this contract is terminated as provided in paragraph (a) hereof, the Government shall be entitled (i) to pursue the same remedies against the Architect-Engineer as it could pursue in the event of a breach of the contract by the Architect-Engineer, and (ii) as a penalty in addition to any other damages to which it may be entitled by law, to exemplary damages in an amount (as determined by the Secretary or his duly authorized representatives) which shall be not less than three nor more than ten times the costs incurred by the Architect-Engineer in providing any such gratuities to any such officer or employee.

(c) The rights and remedies of the Government provided in this clause shall not be exclusive and are in addition to any other rights and remedies provided by law or under this contract.

28. INTEREST (MAY 1963). Notwithstanding any other provision of this contract, unless paid within 30 days all amounts that become payable by the Architect-Engineer to the Government under this contract (net of any applicable tax credit under the Internal Revenue Code) shall bear interest at the rate of six percent per annum from the date due until paid and shall be subject to adjustments as provided by Part 6 of Appendix E of the Armed Services Procurement Regulation, as in effect on the date of this contract. Amounts shall be due upon the earliest one of (i) the date fixed pursuant to this contract, (ii) the date of the first written demand for payment, consistent with this contract, (iii) the date of transmittal by the Government to the Architect-Engineer of a proposed supplemental agreement to confirm completed negotiations fixing the amount of (iv) if this contract provided for revision of prices, the date of written notice to the Architect-Engineer stating the amount of refund payable in connection with a pricing proposal or in connection with a negotiated pricing agreement not confirmed by contract supplement.

29. ACCIDENT PREVENTION (JAN 1965). (a) In order to provide safety controls for protection to the life and health of employees and other persons; for prevention of damage to property, materials, supplies, and equipment; and for avoidance of work interruptions in the performance of this contract, the Architect-Engineer shall comply with all pertinent provisions of Corps of Engineers Mannual, EM385-1-1, dated 13 March 1958, entitled "General Safety Requirements," as amended, and will also take or cause to be taken such additional measures as the Contracting Officer may determine to be reasonably necessary for the purpose.

(b) The Architect-Engineer will maintain an accurate record of, and will report to the Contracting Officer in the manner and on the forms prescribed by the Contracting Officer, exposure data and all accidents resulting in death, traumatic injury, occupantional disease, and damage to property, materials, supplies and equipment incident to work performed under this contract.

(c) The Contracting Officer will notify the Architect-Engineer of any noncompliance with the foregoing provisions and the action to be taken. The Architect-Engineer shall, after receipt of such notice, immediately take corrective action. Such notice, when delivered to the Architect-Engineer or his representative at the site of the work, shall be deemed sufficient for the purpose. If the Architect-Engineer fails or refuses to comply promptly the Contracting Officer may issue an order stopping all or part of the work until satisfactory corrective action has been taken. No part of the time lost due to any such stop orders shall be made the subject of claim for extension of time or for excess costs or damages by the Architect-Engineer.

(d) Compliance with the provisions of this article by subcontractors will be the responsibility of the Architect-Engineer.

30. TERMINATION FOR CONVENIENCE OF THE GOVERNMENT (JAN. 1960). *(This clause is applicable only to contracts $10,000 and under).* The Contracting Officer, by written notice, may terminate this contract, in whole or in part, when it is in the interest of the Government. If this contract is so terminated, the rights, duties and obligations of the parties hereto shall be in accordance with the applicable Sections of the Armed Services Procurement Regulation in effect on the date of this contract.

31. TERMINATION FOR CONVENIENCE OF THE GOVERNMENT (APR. 1966). *(This clause is applicable only to contracts in excess of $10,000).* (a) The performance of

work under this contract may be terminated by the Government in accordance with this clause in whole, or from time to time in part, whenever the Contracting Officer shall determine that such termination is in the best interest of the Government. Any such termination shall be effected by delivery to the Architect-Engineer of a Notice of Termination specifying the extent to which performance of work under the contract is terminated, and the date upon which such termination becomes effective.

(b) After receipt of a Notice of Termination, and except as otherwise directed by the Contracting Officer, the Architect-Engineer shall:

(i) stop work under the contract on the date and to the extent specified in the Notice of Termination;

(ii) Place no further orders or subcontracts for materials, services or facilities, except as may be necessary for completion of such portion of the work under the contract as is not terminated;

(iii) terminate all orders and subcontracts to the extent that they relate to the performance of work terminated by the Notice of Termination;

(iv) assign to the Government, in the manner, at the times and to the extent directed by the Contracting Officer, all of the right, title, and interest of the Architect-Engineer under the orders and subcontracts so terminated, in which case the Government shall have the right, in its discretion, to settle or pay any or all claims arising out of the termination of such orders and subcontracts;

(v) settle all outstanding liabilities and all claims arising out of such termination of orders and subcontracts, with the approval or ratification of the Contracting Officer, to the extent he may require which approval or ratification shall be final for all the purposes of this clause;

(vi) transfer title and deliver to the Government, in the manner, at the times, and to the extent, if any, directed by the Contracting Officer, (A) the fabricated or unfabricated parts, work in process, completed work, supplies, and other material produced as a part of, or acquired in connection with the performance of, the work terminated by the Notice of Termination, and (B) the completed or partially completed plans, drawings, information, and other property which, if the contract had been completed, would have been required to be furnished to the Government.

(vii) use his best efforts to sell, in the manner, at the times, to the extent, and at the price or prices directed or authorized by the Contracting Officer, any property of the types referred to in (vi) above; provided, however, that the Architect-Engineer (A) shall not be required to extend credit to any purchaser, and (B) may acquire any such property under the conditions prescribed by and at a price or prices approved by the Contracting Officer; and provided further that the proceeds of any such transfer or disposition shall be applied in reduction of any payments to be made by the Government to the Architect-Engineer under this contract or shall otherwise be credited to the price of cost of the work covered by this contract or paid in such other manner as the Contracting Officer may direct;

(viii) complete performance of such part of the work as shall not have been terminated by the Notice of Termination; and

(ix) take such action as may be necessary, or as the Contracting Officer may direct, for the protection and preservation of the property related to this contract which is in the possession of the Architect-Engineer and in which the Government has or may acquire an interest.

At any time after expiration of the plant clearance period, as defined in Section VIII, Armed Services Procurement Regulation, as it may be amended from time to time, the Architect-Engineer may submit to the Contracting Officer a list, certified as to quantity and quality, of any or all items of termination inventory not previously disposed of, exclusive of items the disposition of which has been directed or authorized by the Contracting Officer, and may request the Government to remove such items or enter into a storage agreement covering them. Not later than fifteen (15) days thereafter, the Government will accept title to such items and remove them or enter into a storage agreement covering the same; provided, that the list submitted shall be subject to verification by the Contracting Officer upon removal of the items, or if the items are stored, within forty-five (45) days from the date of submission of the list, and any necessary adjustment to correct the list as submitted shall be made prior to final settlement.

(c) After receipt of a Notice of Termination, the Architect-Engineer shall submit to the Contracting Officer his termination claim, in the form and with certification prescribed by the Contracting Officer. Such claim shall be submitted promptly but in no event later than one year from the effective date of termination, unless one or more extensions in writing are granted by the Contracting Officer, upon request of the Architect-Engineer made in writing within such one year period or authorized extension thereof. However, if the Contracting Officer determined that the facts justify such action, he may receive and act upon any such termination claim at any time after such one year period or any extension thereof. Upon failure of the Architect-Engineer to submit his termination claim within the time allowed, the Contracting Officer may, subject to any Settlement Review Board approvals required by Section VIII of the Armed Services Procurement Regulation in effect as of the date of execution of this contract, determine on the basis of information available to him, the amount, if any, due to the Architect-Engineer by reason of the termination and shall thereupon pay to the Architect-Engineer the amount so determined.;

(d) Subject to the provisions of paragraph (c) and subject to any Settlement Review Board approvals required by Section VIII of the Armed Services Procurement Regulation in effect as of the date of execution of this contract, the Architect-Engineer and the Contracting Officer may agree upon the whole or any part of the amount or amounts to be paid to the Architect-Engineer by reason of the total or partial termination of work pursuant to this clause, which amount or amounts may include a reasonable allowance for profit on work done; provided, that such agreed amount or amounts exclusive of settlement costs, shall not exceed the total contract price as reduced by the amount of payments otherwise made and as further reduced by the contract price of work not terminated. The contract shall be amended accordingly, and the Architect-Engineer shall be paid the agreed amount. Nothing in paragraph (e) of this clause prescribing the amount to be paid to the Architect-Engineer in the event of failure of the Architect-Engineer and the Contracting Officer to agree upon the whole amount to be paid to the Architect-Engineer by reason of the termination of work pursuant to this clause, shall be deemed to limit, restrict, or otherwise determine or affect the amount or amounts which may be agreed upon to be paid to the Architect-Engineer pursuant to this paragraph (d).

(e) In the event of the failure of the Architect-Engineer and the Contracting Officer to agree as provided in paragraph (d), upon the whole amount to be paid to the Architect-Engineer by reason of the termination of work pursuant to this clause, the Contracting Officer shall, subject to any Settlement Review Board approvals required by Section VIII of the Armed Services Procurement Regulation in effect as of the date of execution of this contract, pay to the Architect-Engineer the amounts determined by the Contracting Officer as follows, but without duplication of any amounts agreed upon in accordance with paragraph (d):

(i) for completed work and services accepted by the Government, the price or prices specified in the contract for such work, less any payments previously made;

(ii) the total of—

(A) the costs incurred in the performance of the work and services terminated, including initial costs and preparatory expenses allocable thereto, but exclusive of any costs attributable to the work and services paid or to be paid for under paragraph (e) (i) hereof;

(B) the cost of settling and paying claims arising out of the termination of work or services under subcontracts or orders as provided in paragraph (b) (v) above, which are properly chargeable to the terminated portion of the contract (exclusive of amounts paid or payable on account of work or services delivered or furnished by subcontractors prior to the effective date of termination, which amounts shall be included in the costs payable under (A) above); and

(C) a sum, as profit on (A) above, determined by the Contracting Officer pursuant to 8—303 of the Armed Services Procurement Regulation, in effect as of the date of execution of this contract, to be fair and reasonable; *provided,* that if it appears that the Architect-Engineer would have sustained a loss on the entire contract had it been completed, no profit shall be included or allowed under this subdivision (C) and an appropriate adjustment shall be made reducing the amount of settlement to reflect the indicated rate of loss; and

(iii) the reasonable cost of the preservation and protection of property incurred pursuant to paragraph (b) (ix); and any other reasonable cost incidental to the termination of work under this contract, including expense incidental to the determination of the amount due to the Architect-Engineer as a result of the termination of work under this contract.

The total sum to be paid to the Architect-Engineer under (i) and (ii) above shall not exceed the total contract price as reduced by the amount of payments otherwise made and as further reduced by the contract price of work not terminated. Except for normal spoilage, and except to the extent that the Government shall have otherwise expressly assumed the risk of loss, there shall be excluded from the amounts payable to the Architect-Engineer under (ii) above, the fair value as determined by the Contracting Officer of property which is destroyed, lost, stolen, or damaged so as to become undeliverable to the Government, or a buyer pursuant to paragraph (b) (vii).

(f) Any determination of costs under paragraph (c) or (e) hereof shall be governed by the principles for consideration of costs set forth in Section XV, Part 2, of the Armed Services Procurement Regulation, as in effect on the date of this contract.

(g) The Architect-Engineer shall have the right of appeal, under the clause of this contract entitled "Disputes," from any determination made by the Contracting Officer under paragraph (c) or (e) above, except that if the Architect-Engineer has failed to submit his claim within the time provided in paragraph (c) above and has failed to request extension of such time, he shall have no such right of appeal. In any case where the Contracting Officer has made a determination of the amount due under paragraph (c) or (e) above, the Government shall pay to the Architect-Engineer the following: (i) if there is no right of appeal hereunder or if no timely appeal has been taken, the amount so determined by the Contracting Officer, or (ii) if an appeal has been taken, the amount finally determined on such appeal.

(h) In arriving at the amount due the Architect-Engineer under this clause there shall be deducted (i) all unliquidated advance or other payments on account theretofore made to the Architect-Engineer, applicable to the terminated portion of this contract, (ii) any claim which the Government may have against the Architect-Engineer in connection with his contract, and (iii) the agreed price for, or the proceeds of sale of, any materials, supplies, or other things acquired by the Architect-Engineer or sold, pursuant to the provisions of this clause, and not otherwise recovered by or credited to the Government.

(i) If the termination hereunder be partial, prior to the settlement of the terminated portion of this contract, the Architect-Engineer may file with the Contracting Officer a request in writing for an equitable adjustment of the price or prices specified in the contract relating to the contained portion of the contract (the portion not terminated by the Notice of Termination), and such equitable adjustment as may be agreed upon shall be made in such price or prices.

(j) The Government may from time to time, under such terms and conditions as it may prescribe, make partial payments and payments on account against costs incurred by the Architect-Engineer in connection with the terminated portion of this contract whenever in the opinion of the Contracting Officer the aggregate of such payments shall be within the amount to which the Architect-Engineer will be entitled hereunder. If the total of such payments is in excess of the amount finally agreed or determined to be due under this clause, such excess shall be payable by the Architect-Engineer to the Government upon demand, together with interest computed at the rate of 6 percent per annum, for the period from the date such excess payment is received by the Architect-Engineer to the date on which such excess is repaid to the Government; provided, however, that no interest shall be charged with respect to any such excess payment attributable to a reduction in the Architect-Engineer's claim by reason of retention or other disposition of termination inventory until ten days after the date of such retention or disposition, or such later date as determined by the Contracting Officer by reason of the circumstances.

(k) Unless otherwise provided for in this contract, or by applicable statute, the Architect-Engineer from the effective date of termination and for a period of three years after final settlement under this contract, shall preserve and make available to the Government at all reasonable times at the office of the Architect-Engineer but without direct charge to the Government, all his books, records, documents, and other evidence bearing on the costs and expenses of the Architect-Engineer under this contract and relating to the work terminated hereunder, or, to the extent approved by the Contracting Officer, photographs, micro-photographs, or other authentic reproductions thereof.

32. SUBCONTRACTORS AND OUTSIDE ASSOCIATES AND CONSULTANTS (JAN. 1965). Any subcontractors and outside associates or consultant required by the Architect-Engineer in connection with the services covered by the contract will be limited to such individuals or firms as were specifically identified and agreed to during negotiations. Any substitution in such subcontractors, associates, or consultants will be subject to the prior approval of the Contracting Officer.

33. COMPOSITION OF ARCHITECT-ENGINEER (JAN. 1965). If the Architect-Engineer hereunder is comprised of more than one legal entity, each such entity shall be jointly and severally liable hereunder.

34. AUDIT (SEP. 1964). (a) For purposes of verifying that cost or pricing data submitted, in conjunction with the negotiation of this contract or any contract change or other modification involving an amount in excess of $100,000 are accurate, complete, and current, the Contracting Officer, or his authorized representatives, shall-until the expiration of three years from the date of final payment under this contract–have the right to examine those books, records, documents and other supporting data which will permit adequate evaluation of the cost or pricing data submitted, along with the computations and projections used therein, which were available to the Architect-Engineer as of the date of execution of the Architect-Engineer's Certificate of Current Cost or Pricing Data.

(b) The Architect-Engineer agrees to insert the substance of this clause including this paragraph (b) in all subcontracts hereunder in excess of $100,000 so as to apply until three years after final payment under the subcontract, unless the price is based on adequate price competition, established catalog or market prices of commercial items sold in substantial quantities to the general public, or prices set by law or regulation. In each such excepted subcontract hereunder in excess of $100,000, the Architect-Engineer shall insert the substance of the following clause to apply until three years after final payment under the subcontract.

AUDIT-PRICE ADJUSTMENTS. (a) This clause shall become operative only with respect to any change or other modification made pursuant to one or more provisions of this contract which involves a price adjustment in excess of $100,000 unless the price adjustment is based on adequate price competition, established catalog or market prices of commercial items sold in substantial quantities to the general public, or prices set by law or regulation and further provided that such change or other modification to this contract must result from a change or other modification to the Government prime contract.

(b) For purposes of verifying that cost or pricing data submitted in conjunction with a contract change or modification involving an amount in excess of $100,000 are accurate, complete and current, the Contracting Officer or his authorized representative shall–until the expiration of three years from the data of final payment under this contract–have the right to examine those books, records, documents, and other supporting data which will permit adequate evaluation of the cost or pricing data submitted, along with the computations and projections used therein, which were available to the Contractor as of the date of execution of the Contractor's Certificate of Current Cost or Pricing Data.

(c) The Contractor agrees to insert the substance of this clause, including this paragraph (c), in all subcontracts hereunder in excess of $100,000 so as to apply until three years after final payment of the subcontract.

35. SUBCONTRACTOR COST AND PRICING DATA (SEP. 1964). (a) The Architect-Engineer shall require subcontractors hereunder to submit cost or pricing data under the following circumstances: (i) prior to award of any cost-reimbursement type, incentive, or price redeterminable subcontract; (ii) prior to the award of any subcontract the price of which is expected to exceed $100,000; (iii) prior to the pricing of any subcontract change or other modification for which the price adjustment is expected to exceed $100,000; except in the case of (ii) or (iii) where the price is based on adequate price competition, established catalog or market prices of commercial items sold in substantial quantities to the general public, or prices set by law or regulation.

(b) The Architect-Engineer shall require subcontractors to certify, in substantially the same form as that used in the certificate by the Architect-Engineer to the Government, that to the best of their knowledge and belief, the cost and pricing data submitted under (a) above is accurate, complete, and current as of the date of execution, which date shall be as close as possible to the date of agreement on the negotiated price of the subcontract.

(c) The Architect-Engineer shall insert the substance of this clause including this paragraph (c) in each of his cost-reimbursement type, price redeterminable, or incentive subcontracts hereunder, and in any other subcontract hereunder which exceeds $100,000 except where the price thereof is based on adequate price competition, established catalog or market prices of commercial items sold in substantial quantities to the general public, or prices set by law or regulation. In each such excepted subcontract hereunder in excess of $100,000, the Architect-Engineer shall insert the substance of the following clause:

SUBCONTRACTOR COST AND PRICING DATA-PRICE ADJUSTMENTS. (a) Paragraphs (b) and (c) of this clause shall become operative only with respect to any change or other modification made pursuant to one or more provisions of this contract which involves a price adjustment in excess of $100,000. The requirements of this clause shall be limited to such price adjustments.

(b) The Contractor shall require subcontractors hereunder to submit cost or pricing data under the following circumstances: (i) prior to award of any cost-reimbursement type, incentive or price redeterminable subcontract; (ii) prior to award of any subcontract, the price of which is expected to exceed $100,000; (iii) prior to the pricing of any subcontract change or other modification for which the price adjustment is expected to exceed $100,000; except, in the case of (ii) or (iii), where the price is based on adequate price competition, established catalog or market prices of commercial items sold in substantial quantities to the general public, or prices set by law or regulation.

(c) The Contractor shall require subcontractors to certify, in substantially the same form as that used in the certificate by the Prime Contractor to the Government, that to the best of their knowledge and belief the cost and pricing data submitted under (b) above is accurate, complete, and current as of the date of execution, which date shall be as close as possible to the date of agreement of the negotiated price of the contract modification,

(d) The Contractor shall insert the substance of this clause including this paragraph (d) in each subcontract hereunder which exceeds $100,000;

36. PRICE REDUCTION FOR DEFECTIVE COST OR PRICING DATA (SEP. 1964). (a) If the Contracting Officer determines that any price, including profit or fee, negotiated in connection with this contract was increased by any significant sums because the Architect-Engineer, or any subcontractor in connection with a subcontract covered by (c) below, furnished incomplete or inaccurate cost or pricing data or data not current as certified in the Architect-Engineer's Certificate of Current Cost or Pricing Data, then such price shall be reduced accordingly and the contract shall be modified in writing to reflect such adjustment.

(b) Failure to agree on a reduction shall be a dispute concerning a question of fact within the meaning of the "Disputes" clause of this contract.

(c) The Architect-Engineer agrees to insert the substance of paragraphs (a) and (c) of this clause in each of his cost-reimbursement type, price redeterminable, or incentive subcontracts hereunder, and in any other subcontract hereunder in excess of $100,000 unless the price is based on adequate price competition, established catalog or market prices of commercial items sold in substantial quantities to the general public, or prices set by law or regulation. In each such excepted subcontract hereunder which exceeds $100,000, the Architect-Engineer shall insert the substance of the following clause.

PRICE REDUCTION FOR DEFECTIVE COST OR PRICING DATA-PRICE AD-JUSTMENTS. (a) This clause shall become operative only with respect to any change or other modification made pursuant to one or more provisions of this contract which involves a price adjustment in excess of $100,000, except where the price is based on adequate price competition, established catalog or market prices of commercial items sold in substantial quantities to the general public, or prices set by law or regulation. The right to price reduction under this clause shall be limited to such price adjustments.

(b) If the Contractor determines that any price, including profit or fee, negotiated in connection with any price adjustment within the purview of paragraph (a) above was increased by any significant sum because the subcontractor or any of subcontractors in connection with a subcontract covered by paragraph (c) below, furnished incomplete or inaccurate cost or pricing data, or data not current as of the date of execution of the subcontractor's certificate of current cost or pricing data, then such price shall be reduced accordingly and the subcontract shall be modified in writing to reflect such adjustment.

(c) The subcontractor agrees to insert the substance of this clause in each subcontract hereunder which exceeds $100,000.

37. EQUAL OPPORTUNITY. (a) The "Equal Opportunity" clause in paragraph 24 is amended by deleting references to the President's Committee on Equal Employment Opportunity, Executive Order 10925 of March 6, 1961, as amended, and Section 303 of Executive Order No. 10925 of March 6, 1961, as amended, and substituting therfor the Secretary of Labor, Executive Order No. 11246 of September 24, 1965, and Section 204 of Executive Order 11246 of September 24, 1965, respectively.

(b) The Equal Opportunity representation in paragraph 24 is amended to insert after the reference to "Executive Order 10925" the following: "or the clause contained in Section 201 of Executive Order No. 11114."

(c) In accordance with regulations of the Secretary of Labor, the rules, regulations, orders, instructions, designations, and other directives issued by the President's Committee on Equal Employment Opportunity and those issued by the heads of various departments or agencies under or pursuant to any of the Executive Orders superseded by Executive Order 11246, shall, to the extent that they are not inconsistent with Executive Order 11246, remain in full force and effect unless and until revoked or superseded by appropriate authority. References in such directives to provisions of the superseded orders shall be deemed to be references to the comparable provisions of Executive Order 11246.

STANDARD CONTRACT FORMS

The American Institute of Architects publishes two other standard forms of agreement between owner and architect in addition to Document B131 reproduced in this chapter. The full series consists of:

AIA Document B131 – Owner-Architect Agreement: Percentage of Construction Cost.
AIA Document B231 – Owner-Architect Agreement: Multiple of Direct Personnel Expense.
AIA Document B331 – Owner-Architect Agreement: Professional Fee Plus Expense.

The AIA also publishes a matching series of standard forms of agreement covering consulting services for use by architectural firms operating without in-house engineering capability. Of the three forms listed below, the first is reproduced at the end of Chapter 3.

AIA Document C131 – Architect-Engineer Agreement: Percentage of Construction Cost.
AIA Document C231 – Architect-Engineer Agreement: Multiple of Direct Personnel Expense.
AIA Document C231 – Architect-Engineer Agreement: Professional Fee Plus Expense.

Chapter 3

Architect-Consultant Contracts

Adequacy of Consulting Engineering Agreements

Although an architectural firm may retain the services of consultants in fields other than structural, mechanical, and electrical engineering, the bulk of consulting services utilized in current practice lies in these three areas. Thus the discussion that follows is focused on the contractural relationships between architects and consulting engineers and consideration of architect-consultant contracts in other fields such as acoustics, lighting, landscape architecture, etc., is deferred until later in the chapter.

Since the scope of owner-architect contracts for design services includes all normal structural, mechanical, and electrical engineering required for a building project, architectural firms that do not possess in-house engineering capability secure it by retaining consultants. The agreement between architect and consulting engineer provides for the consulting engineer to assume some of the duties undertaken by the architect in his contract with the owner; however, the architect does not thereby divest himself of responsibility for the consultant's part of the project. If consulting engineers were retained directly by the owner, the architect would not be responsible for their performance; the architect, however, could hardly carry out his coordinating functions effectively under such conditions and the prime advantage of single design responsibility would be lost to the owner. There is no practicable way for the architect to be relieved of responsibility for the work of consultants and consequently he must accept this responsibility even though he is not for the most part competent to verify their work. It is therefore essential that contracts with consulting engineers be drawn so that the engineer is legally responsible to the architect for his part of the project.

On the other hand, it should also be made clear that the consulting engineer's responsibility does not extend beyond that portion of the work he was retained to do. The contract should state that the engineer, as an independent contractor, is engaged by the architect as his agent for a specific part of the project. The intention here is to avoid implication that architect and consultant are engaged in a joint venture where both parties usually have full joint responsibility for the other's actions. In order to reinforce the agency concept as against joint venture, the architect's obligation to pay the consulting engineer should not be made contingent on his receiving his own fee from the owner. However, timing of payments may be made the same as that provided by the schedule in the owner-architect contract (that is, the prime professional contract).

Even with the greatest care it is very difficult to delineate clear lines of responsibility between architect and consulting engineer and to limit the rights of the owner with respect to either. The standard forms of agreement between architect and engineer promulgated by the American Institute of Architects, the Consulting Engineers Council, and the National Society of Professional Engineers have been very carefully drawn in this regard, but it must be recognized that a suit can be brought against both architect and consultant by the owner, the construction contractor, or third parties if anyone wishes to. This situation makes it prudent for both consulting engineer and architect to carry professional liability insurance.

The architect-consultant contract should state clearly that the architect as prime professional contractor shall be the project coordinator and that the consultant shall collaborate with him in furnishing professional services for the part of the project concerned. It should also state that the consulting engineer's duties and responsibilities with respect to this part of the project shall be co-extensive with those of the architect to the owner under the prime contract. Since the consulting engineer does, in fact, perform some of the services for which the owner contracts with the architect, it is necessary that the engineer be furnished a copy of the owner-architect contract, or at least a copy of the pertinent portions so that these may be made a part of the consulting engineering agreement by reference. Just what may constitute the "pertinent portions" of the prime contract will vary with circumstances, but presumably the provisions represented by the *Terms and Conditions* of AIA Document B131 (discussed in the preceding chapter) plus those indicating the basis and timing of payments, although not the amount of compensation, would represent a minimum.

It is also essential that the intent and language of the owner-architect and architect-consultant agreements be coordinated. In order to emphasize the relationship between these agreements, the articles in the discussion that follows are arranged in the same order as corresponding topics in the preceding chapter.

Standard contract forms have been included at the end of the chapter for reference and comparative study.[1]

Types of Consulting Engineering Contracts

The principal types of architect-consultant contracts for engineering services as related to basis of compensation are

Percentage of construction cost of consultant's part of the project
Lump-sum consulting fee
Multiple of consultant's direct personnel expense
Professional consulting fee plus expense

In general, it will be advantageous on any given project to employ the same type of contract for consulting engineering services as that used for the owner-architect agreement. When circumstances seem to indicate otherwise, caution should be exercised in selecting the type of consulting contract. Although there is no disadvantage from the architect's point of view in making lump-sum agreements with consultants when the prime agreement is of the fee-plus-expense or other expense-reimbursement type, the reverse situation presents some hazards. These may not be significant when the extent of consulting services is known to be small with respect to total services to be performed by the architect, but under certain conditions such as "creeping" increase in scope of the project the architect may find himself obligated to pay for expanding engineering services while he is still working under a lump-sum with the owner. When both prime contract and consultant contract are drawn on the same basis, such risks are at least shared by both professionals involved pending possible renegotiation of the prime contract fee.

Of course, it is questionable whether a consulting engineer would be acting prudently in accepting a lump-sum architect-consultant contract where the prime agreement was based on one of the expense-reimbursement types. As noted in Chapter 2, both professional-fee-plus-expense and multiple-of-direct-personnel-expense types of owner-architect agreements are designed for situations where the scope of A & E services required cannot be determined with a reasonable degree of precision at the inception of a project. Under these circumstances the scope of engineering services presumably would be also too vague to warrant a lump-sum consulting agreement.

Substantially all of the discussion in Chapter 2 concerning relative merits of the several types of owner-architect contracts applies to corresponding types when used for consulting engineering agreements, with the architect in the role

---

[1]These are American Institute of Architects Document C131; Document AE.1 issued by the Consulting Engineers Council; California Council, AIA Document No. 105-65.

of "client" with respect to consulting engineer. Additional factors governing selection and operation of the different types, with particular reference to matters concerning consultant's compensation, are discussed in the following article.

Aspects of Fee Determination

In the areas of "normal" structural, mechanical, and electrical engineering which comprise the bulk of engineering services related to building projects, a percentage of the construction cost of work designed and specified by the engineer is widely used as a basis for establishing consulting fees. Here, as was similarly noted concerning owner-architect contracts, recommended percentage fee schedules promulgated by the various professional societies concerned may be used with confidence as a basis for negotiating architect-consultant contracts. In addition to their obvious relevance to percentage-of-construction-cost contracts, such schedules are also useful for checking lump-sum proposals and estimates of total cost of services contained in proposals based on fee-plus-expense or multiple-of-direct-personnel-expense types of agreements.

An alternative basis for fee determination frequently used for structural engineering work is a percentage of the construction cost of the entire project. The use of project construction cost (rather than construction cost of that part of the project designed by the structural engineer) is predicated on experience indicating that the cost of the structural system of a building is not normally subject to as much change during development of the project as are the mechanical and electrical systems.

Another method used successfully by some firms is establishing the consulting engineer's compensation as a percentage of the architect's compensation instead of a percentage of the construction cost. When the same architect and consulting engineer have been working together over a long period and when the work consists of several projects of the same building type, this kind of agreement may be the simplest arrangement. Under other circumstances, however, it should be used with caution.

The preceding general comments with respect to types of consulting contracts and fee determination are supplemented below by discussion related to specific types of agreements. For easier reference and comparison the order of presentation corresponds to the types-of-contracts discussion in Chapter 2.

PERCENTAGE OF CONSTRUCTION COST OF CONSULTANT'S PART OF THE PROJECT

When the prime A & E contract is of the percentage-of-construction-cost type, this basis of compensation is equitable to both consulting engineer and architect. Compensation to the engineer and costs to the architect are in direct proportion to the work performed. It is the simplest form of contract to operate when the design budget is controlled so that an acceptable low bid may be obtained within the limit of cost imposed.

LUMP-SUM CONSULTING FEE

Here, as in owner-architect agreements, the lump-sum may be based on a percentage of the estimated cost of the consultant's part of the project or on the consultant's estimated man-hour costs of performing his work. The principal problem with the first agreement is that of arriving at a reasonably accurate estimate of the construction cost of the work involved. In arriving at a lump-sum fee for the prime professional contract the construction cost of the project as a whole is known from the limit set in the owner's budget. However, the limit of cost for the mechanical (or electrical) engineer's part of the project usually is not known as precisely at the time consulting services are negotiated. Although order-of-magnitude estimates can be made on the basis of previous experience, it commonly happens in practice that, for example, the quality and sophistication of air conditioning systems develop and change as project design proceeds and budgetary allocations in the building program are adjusted.

Because of this situation some mechanism should be provided in the consulting contract so that the engineer does not have to make a "final firm" commitment for a lump-sum so early in the project development and on insufficient information. If such provision is not included and the scope of the engineer's work is expanded (within the same overall scope of the project), he would be penalized by having to design more work than he originally intended. On the other hand, if the scope of the engineer's work were adjusted downward, the architect would be overpaying for the consulting services at the expense of his own project budget. To compensate for this situation a statement may be incorporated in the contract similar to the following:

"It is further agreed that the lump-sum of $00,000, which is 0.0% of the tentatively estimated $0,000,000 construction cost of This Part of the Project, will be adjusted to conform to the amount resulting from this same percentage of the estimate for this Part of the Project that is subsequently agreed upon at the end of the Design Development Phase."[2]

Note that this procedure permits an adjustment of the lump-sum fee on the basis of a subsequent *estimate* only. To make an adjustment on the basis of bids received would circumvent the purposes of this type of contract.

MULTIPLE OF CONSULTANT'S DIRECT PERSONNEL EXPENSE

This is the simplest type of contract for use when payments to the consulting

[2]Under some circumstances it may be preferable to substitute for "Design Development Phase" the phrase "Construction Document Phase prior to calling for bids on the project."

engineer are to be based on cost of services rendered rather than on a percentage or lump-sum fee. Precautions with reference to its use under a lump-sum prime A & E contract were discussed at the beginning of this article. However, when circumstances related to a project make the multiple-of-direct-personnel-expense type agreement appropriate for the prime A & E contract, it is both convenient and equitable to use this type for consulting agreements. The multiple used by the consulting engineer usually approximates that of the architect and in any event it must be acceptable to the owner since it is common practice to stipulate the multiple for consultant's direct personnel expense in the owner-architect agreement. Furthermore, parallel to provisions in the prime contract, if participation of the consulting firm's principals is not accounted for in the multiplier, a specific statement in the consulting agreement should stipulate a fixed rate for principals' time and should identify the persons so designated. Obviously, agreements of this type require that the accounting system of the consulting engineering firm be adequate and available for audit by the architect and/or owner.

PROFESSIONAL CONSULTING FEE PLUS EXPENSE

As employed for architect-consultant contracts, the "expense" in this type of agreement is generally understood to be a multiple of the consulting engineer's direct personnel expense plus "reimbursable" expenses similar to those provided for in the prime A & E contract. The fixed professional fee represents the firm's engineering skill and responsibility, participation of the principals, and availability of its resources. Like its counterpart in owner-architect contracts, this type of agreement results in a professional fee that is independent of final design costs unless the scope of the consultant's part of the work is changed.

When the prime A & E contract between owner and architect is based on a professional fee plus expense with guaranteed ceiling, the architect-consulting engineer agreement should also have a ceiling guaranteed by the engineer. This type of arrangement can become quite cumbersome to operate when several consultants are involved and when changes are made in the scope of the project, thereby necessitating renegotiation of the several fixed fees and guaranteed ceilings. One device used by architects to simplify administrative procedures under a prime A & E contract with a guaranteed ceiling is to negotiate initial agreements with consultants on the basis of multiple of direct personnel expense. These are subsequently converted to lump-sum agreements during the schematic design phase or early in the design development phase when project characteristics have become sufficiently definitive.

Scope of Consulting Services

The scope of work to be performed by the consulting engineer in connection with his part of the project is, of course, the same as that which would be accomplished by the architectural firm if it possessed in-house engineering capability. This is why, as mentioned earlier, the consulting contract should state explicitly that the engineer's duties and responsibilities are co-extensive with those of architect to owner under the prime agreement as related to the consulting engineer's part of the project.

In parallel with the discussion of scope of professional services in Chapter 2, the scope of consulting services before and after award of the construction contract are considered separately; also the discussion is keyed to percentage-of-construction-cost or lump-sum types of consulting contracts. The functions to be performed, however, are the same under cost-reimbursement-type contracts, the principal difference being lack of significance between basic services and "additional services."

Consulting Services Prior to Award of Construction Contract

In addition to engineering consultation that may be required in programming or other project analysis services, the consulting engineer's participation before the award of the construction contract is executed within the framework of the first four phases of basic A & E services. Consequently, the consulting contract should indicate the services to be undertaken by the engineer during schematic design, design development, construction document, and bidding phases. Although local custom may designate these phases by different names and occasionally combine some of them, the total scope of services required must be performed.

SCHEMATIC DESIGN PHASE

Since this phase of the project is intended to set the general functional scheme with respect to previously determined space requirements and relationships, occupancy requirements, and budget limitations, the consulting agreement should call for the engineer to prepare necessary analyses of different systems that might be used, furnish an estimate of the probable construction cost or range of cost for his part of the project, and attend necessary conferences in connection with preparing the schematic design.

DESIGN DEVELOPMENT PHASE

During this phase the principal quantitative engineering problems are worked out. These include approximate space requirements for structural, mechanical, and electrical systems; capacities and performance characteristics of equipment; and a more refined estimate of the construction cost of the engineer's part of the project. The consulting agreement should call for the engineer to prepare design development drawings and outline specifications in conformance with the architect's design, delineating the kinds and quality of materials and equipment contemplated.

CONSTRUCTION DOCUMENT PHASE

The consulting agreement should stipulate that, based on the architect's notice of approval of design development documents, the engineer shall prepare working drawings and specifications completely delineating the work required for his part of the project. In addition, the engineer shall make any adjustments in his previous cost estimates that may result from changes in scope, requirements, or market conditions. The provisions of this phase sometimes place on the engineer the responsibility for filing required drawings and other documents to obtain approval of governmental agencies exercising jurisdiction over the design of his part of the project.

BIDDING OR NEGOTIATION PHASE

The principal point here is to include provision that the engineer shall assist the architect in analyzing bids, or proposals otherwise submitted, for construction of this part of the project.

Consulting Services Subsequent to Award of Construction Contract

Since the consulting engineer's duties and responsibilities for his part of the work are co-extensive with those of the architect to the owner, the scope of consulting services to be performed during construction will depend on whether the prime A & E contract calls for full or abridged construction phase services as described in Chapter 2. The discussion here relates to full services.

ENGINEER'S AUTHORITY

In general, the intent of architect-consultant agreements is to make certain that the engineer shall assist the architect in making decisions on all claims of the owner and construction contractor that relate to the engineer's part of the project. Unless otherwise specifically provided, however, the consulting engineer will not render final decision on claims between owner and contractor, since this responsibility is customarily placed on the architect by the terms of the construction contract between owner and contractor. If this situation is modified so that the consulting engineer is to have direct contacts with owner, construction contractor, and subcontractors (instead of working through the architect), the extent of the engineer's authority must be made clear to all concerned in addition to being spelled out in the consulting contract. It should also be borne in mind that there is some uncertainty concerning the degree to which working agreements between architect and consultants delineating their responsibilities and authorities can be made binding on owner and construction contractor.

SUPERVISION OF CONSTRUCTION

The same precaution with respect to use of the term "supervision" in prime

A & E contracts applies to consulting agreements. Here again, the real purpose of the consulting engineer's on-site visits and observations is to "... endeavor to guard ..." against defects and deficiencies, not to *supervise* the construction contractor's work. The consulting agreement should provide that the engineer shall inform the architect of any defects or deficiencies relating to construction of his part of the project and make recommendations to the architect concerning action to be taken where work fails to conform with requirements of the drawings and specifications. No more than does the architect, however, does the consulting engineer guarantee the work of the construction contractor for this part of the project.

PAYMENT CERTIFICATES

Provision should be made for the consulting engineer to assist in evaluating the reasonableness of requests for payment submitted by the construction contractor. Since the architect is required to issue certificates of payment over his signature, it is common practice to require that the engineer certify to the architect amounts to be included in payment certificate items that are related to construction of his part of the project. By certifying such amounts as owing to the contractor the consulting engineer represents that construction of the part of the project concerned has progressed to the extent indicated and that to the best of his knowledge and belief it is in accordance with the drawings and specifications.

CHECKING OF SHOP DRAWINGS

The checking of shop drawings by the consulting engineer follows the same general pattern as that provided in the prime professional contract, that is, checking only for conformance with the design concept of this part of the project and compliance with plans and specifications relating to them. Here again approvals should never imply responsibility for accuracy of shop drawings, especially in matters relating to fabrication and construction techniques which are outside the professional competency of the consultant. The same provisions apply generally to approval of samples, schedules, performance charts for equipment, and other submissions by the construction contractor. It should also be stated in the consulting agreement that all the consulting engineer's comments and approvals with respect to shop drawings and other submissions shall be routed through the architect as project coordinator.

Consultant's Services Not Included

In the interest of establishing a clear understanding of work to be performed by the consulting engineer, specific items not included in basic consulting services should be so identified. These fall under the two general headings of information to be furnished by the architect and items considered as additional services for which the engineer will be reimbursed separately.

ARCHITECT-FURNISHED INFORMATION

It is generally accepted practice to stipulate that the architect shall provide full information concerning requirements for the consultant's part of the project. In addition, he shall furnish a copy of the owner's certified survey of the site giving information pertinent to this part of the project. This normally includes location, availability, and nature of service and utility lines related to design of the mechanical and electrical systems as well as soils information for use in foundation design and in appraising the feasibility of different structural systems.

ADDITIONAL CONSULTING SERVICES

The pattern here should parallel that embodied in the prime A & E contract relating to the nature of the additional services and method of payment for them by the architect. In this connection a comparison of the article on additional services in AIA DOC. B131 (end of Chapter 2) with those contained in the standard forms for consulting contracts given at the end of this chapter will be of interest. However, the question of scope of technical services to be performed by the consultant deserves most careful consideration. The consulting agreement should be specific on points such as whether the mechanical engineering consultant is to be responsible for design of all conveyor systems in the project, including elevators. The consulting electrical engineer should be clear concerning his responsibility in connection with design of electronic equipment installations, closed circuit television, and other special communication systems. "Normal" engineering design services for different building types may or may not be considered to embrace these and similar items, depending upon local custom. Where there is any chance of conflicting views on these matters it is advisable to establish a list of agreed-upon exceptions to the basic technical areas intended to be encompassed in the consulting agreement. A subsequent decision to include them can then be handled under the provision for additional services.

Construction Cost of Consultant's Part of the Project

As noted in Chapter 2, the term *construction cost* as used in this book is synonomous with *net construction cost.* Consequently, the construction cost of the consulting engineer's part of the project means the cost of all work designed or specified by the engineer, but it does not include the cost of architectural, engineering, or other consulting services. Parallel to provisions of the prime A & E contract, where labor or material is furnished by the owner, their current market cost will be included in the construction cost of this part of the project for purpose of fee determination. In addition, to cover cases where bids are not taken and construction does not proceed, provisions analogous to those for establishing construction cost embodied in the prime contract should be written into the consulting agreement. These normally provide that, in the absence of a bona fide bid, the construction cost of the engineer's part of the project shall be based on the latest detailed cost estimate or on the engineer's latest statement of probable construction cost when a detailed cost estimate has not been made.

Consultant's Cost Estimate Responsibility

The consulting engineer faces the same uncertainties with respect to precision of prebidding estimates covering his part of the project as the architect encounters in forecasting the cost of the project as a whole. Nevertheless, when working under a limit of cost, the architect must rely on his consultants' estimates for the structural, mechanical, and electrical systems, with the consulting engineers assuming some responsibility in this regard. One mechanism used frequently to implement this responsibility is to include provision in the consulting contract that, if the lowest bona-fide bid received exceeds the *engineer's* latest estimate (for his part of the project) by more than a stated percentage, he shall make the necessary revision at his own expense to bring the cost within the agreed-upon range. The permissible overrun of the "latest estimate" is often set at 15%. This figure, however, should be reviewed with reference to the character of individual projects and the provisions of the owner-architect contract relating to the architect's responsibility when no bids are received within the overall project limit of cost.

Payments and Reimbursable Expenses

As noted in the first article of this chapter, the timing of payments between architect and consulting engineer may be made the same as that provided by the schedule for owner's payments to architect contained in the prime professional contract. When the owner-architect agreement (lump-sum or percentage-of-construction-cost type) calls for monthly progress payments covering basic services, it is customary to make the same provision in the consulting engineering contract. Otherwise payments to the engineeer are made at the ends of each phase of the basic services in accordance with an agreed-upon schedule.

Where the multiple-of-direct-personnel-expense type of consulting contract is employed it is customary to provide for monthly payments to the engineer covering current services performed. The same practice exists for the professional consulting fee-plus-expense type as far as direct personnel expense incurred is concerned; but payments on the fixed professional fee are limited to specified percentages related to the phases of basic services.

It is necessary in lump-sum and percentage-of-cost types of consulting contracts to stipulate the method of payment for additional services. The most convenient arrangement is to follow the procedure used in multiple-of-direct-expense type agreements, the method cited in AIA Document C131 given at the end of this chapter. In any event, it is important that billings and payments for the consulting engineer's additional services be segregated from those pertaining to his basic services.

REIMBURSABLE EXPENSES

The language of the consulting contract relating to reimbursable expenses must be coordinated closely with that of the owner-architect agreement. Any restrictions such as prior authorization for travel that may appear in the prime professional contract will normally become a provision of the consulting agreement. For this reason it is usual to state that the engineer's costs for travel, long distance telephone, telegrams, etc., made in the interest of his part of the project will be reimbursed when such expenditures are authorized by the architect. Provision is usually made for payment of the consulting engineer's reimbursable expenses monthly on presentation of a statement covering expenses incurred.

General Terms and Conditions of Consulting Contracts

It should be noted that, just as there are several additional provisions that should form a part of any owner-architect contract for full architectural and engineering services, so there are analogous provisions that should be incorporated in all contracts for consulting services. These are concerned with ownership of plans, specifications, and design calculations; termination of the consulting agreement; settlement of disputes; successors and assigns; extent and accessibility of consultant's accounting records; insurance and mutual assistance. It is recommended that the reader compare articles covering these general provisions in the consulting contract forms at the end of this chapter. A discussion of selected items follows.

Ownership of Plans and Specifications

If all engineering plans and specifications for a project were prepared by an architectural or architect-engineer firm with its own in-house engineer staff, the comments on ownership of these documents given in Chapter 2 would cover the situation. When the prime contract for A & E services is with a government agency that requires a clause vesting ownership of documents in the government, the situation is similarly clear since any contracts the architect might make with consulting engineers would carry the same document ownership clause as the prime professional contract. This would also be true with respect to a private client who insisted on such a clause. In all prime professional contracts that stipulate the "instruments of service" concept, however, the consulting professional has an interest in the ownership (and consequent potential re-use) of the designs, drawings, and specifications he prepares.

Although many of the recommended standard forms for consulting engineering contracts are silent on the ownership of documents question as such, they do contain clauses limiting the *re-use* of the engineer's tracings, master specification sheets, or their reproductions without his written authorization. Although the actual original drawings, tracings, and master specification sheets may become the physical property of the architect, the consulting engineer does not necessarily relinquish his rights in his "design."

As noted in the discussion of this question in Chapter 2, the significance of these provisions will vary with the type of project; and where multiple use of plans and specificatons is contemplated, this may be reflected in the fee structure by establishing re-use "royalties" or similar provisions.

All contracts between architects and consulting engineers should provide that the engineer maintain his design calculations on file in legible form and available to the architect. It is also necessary that calculations be made and maintained so that they are reproducible for reviewing agencies that may require copies.

Termination of Consulting Contract

It is essential that provision be made in architect-consultant contracts so that the agreement will terminate automatically upon termination of the prime professional contract between owner and architect for any cause. In addition, the consulting agreement should provide for termination because of nonperformance of either party. The method of settlement in event of termination should be specified and include procedures for notification of intent to terminate and for establishing the basis of payment to the consulting engineer for work accomplished to date.

Provisions for reimbursing the consulting engineer when termination results from the fault of others differ among the standard forms for consulting contracts. In those issued by the American Institute of Architects, payments to the engineer for terminal expenses are made contingent on comparable adjustment of the architect's compensation with respect to the prime professional contract, whereas the form recommended by the Consulting Engineers Council omits any reference to such contingency.[3]

---

[3]Compare the termination articles in AIA Document C131 and CEC Document AE.1 at the end of this chapter.

Arbitration of Disputes

It is now common practice to provide that disputes arising under architect-consultant contracts be submitted to arbitration. This method, utilizing as it does the judgment of persons knowledgeable in the nature of the question, is particularly appropriate for the resolution of disputes between design professionals. The discussion of arbitration under owner-architect contracts in Chapter 2 is equally applicable to contracts for consulting services.

Insurance and Mutual Assistance

It was pointed out at the beginning of this chapter that although the consulting contract between architect and engineer provides for the engineer to assume some of the duties undertaken by the architect in the prime professional contract, the architect does not thereby divest himself of responsibility for the consultant's part of the project. Consequently, in view of the factors discussed in the article on professional responsibility and liability in Chapter 2, it has become customary to include clauses in consulting contracts whereby each party agrees to furnish "all reasonable assistance" to the other in connection with claims asserted against either relating to performance of professional services.

The current trend by architects is to reinforce this provision by requiring that the consulting engineer carry professional liability insurance. The clauses in the contract covering this point are usually phrased to operate both ways, that is, the architect and engineer are each required to carry insurance that will protect him from claims arising from his own acts, errors, or omissions and those of his employees or others for whom he is legally responsible. To avoid duplication of coverage and consequent inflation of total premium costs, insurance counsel should be consulted when establishing specific provisions.

Nonengineer Consulting Agreements

The complexity as well as the scope of many building projects frequently make it necessary for the architectural firm to retain consulting services in various specialized fields, regardless whether they possess in-house engineering capability in general structural, mechanical, and electrical engineering. Some of these special fields are architectural acoustics and sound transmission control, lighting, color, site planning and landscape architecture, traffic and parking, food handling and kitchen equipment, special electronic communication systems, hardware, elevators, etc. The architect must, of course, foresee the need for such consulting assistance and provide for its cost when establishing his fee for the project, or the client and the architect must agree about which of the "special consultants" the architect may be reimbursed for under the additional services provision.

The owner may, of course, elect to pay for some consulting services directly, which frequently occurs with certain building types such as hospitals and schools that have highly specialized occupancy and functional requirements. There is then only an indirect or cooperative relationship between these consultants and the architect—no direct contractual obligation. Such an arrangement, however, is generally not satisfactory for most of the technical consulting services enumerated in the preceding paragraph if the concept of single design responsibility is to be maintained.

Where the direct relationship between architect and consultant is employed (that is, consultant retained and paid by architect) a consulting agreement similar to the consulting engineering contract is called for. Not all of the specialized consulting services warrant, however, such detailed treatment on every project. In some instances a purchase order will be sufficient; in others a letter-type agreement covering the more significant provisions may be employed. The shorter contract forms such as the California Form of Agreement for Consultant Services (reproduced at the end of this chapter) are especially useful in this connection. Decisions regarding the appropriateness of the use of purchase orders as well as the drafting of letter-type agreements for specific projects should, of course, have the benefit of legal counsel. Similarly, as required for consulting engineering contracts, care must be exercised to see that nonengineering consulting agreements reflect the provisions of the prime professional contract.

Engineer as Prime Professional

Although the situation arises more frequently in connection with heavy constructon, the engineer may be the prime professional services contractor in some types of building projects. When this occurs the architect may be retained as a consultant to collaborate with the engineer who, as prime professional, functions as the project coordinator. Substantially all of the points discussed in this chapter apply with equal validity to this reciprocal relationship between the professionals involved and should be considered when drafting agreements where the architect serves as a consultant instead of in his more usual capacity as prime professional.

EXAMPLES OF CONTRACT FORMS

Three standard forms of architect-consultant contracts are given here as illustrations. The first two relate strictly to consulting engineering services and the third, shorter form is designed for all types of consulting services in situations where greater elaboration than provided for therein is not required. Reproduction is by permission, respectively, of the American Institute of Architects, the Consulting Engineers Council, and the California Council, AIA.

THE AMERICAN INSTITUTE OF ARCHITECTS

AIA Document C131

Standard Form of Agreement Between the Architect and the Engineer

on a basis of a

PERCENTAGE OF CONSTRUCTION COST

AGREEMENT

made this   day of   in the year of Nineteen Hundred and

BETWEEN

the Architect, and

the Engineer.

The Architect has made an agreement dated   with

The Owner, a copy of the Terms and Conditions and any other pertinent portions of which is attached, made a part hereof and marked Exhibit A, and is hereinafter referred to as the Prime Agreement and provides for furnishing professional services in connection with the Project described therein as follows:

The Architect and Engineer agree as set forth below.

AIA DOCUMENT C231 • ARCHITECT-ENGINEER AGREEMENT • SEPTEMBER 1967 EDITION • AIA®
© 1967 THE AMERICAN INSTITUTE OF ARCHITECTS, 1735 N. Y. AVE., N.W., WASH., D. C. 20006

I.   THE ENGINEER shall provide the following professional services for the Architect, in accordance with the Terms and Conditions of this Agreement, which the Architect is required to provide for the Owner under the Prime Agreement:
(Describe services)

The Part of the Project for which the Engineer is to provide such services is hereinafter called This Part of the Project.

II. THE ARCHITECT shall compensate the Engineer, in accordance with the Terms and Conditions of this Agreement, as follows:

   a. *FOR THE ENGINEER'S BASIC SERVICES,* as described in Paragraph 1.2, a Basic Fee of   per cent (   ) of the Construction Cost of the Project, or   per cent (%) of the Construction Cost of This Part of the Project.

   b. *FOR THE ENGINEER'S ADDITIONAL SERVICES* when authorized by the Architect, as described in Paragraph 1.4, a fee computed as follows:

      Principals' time at the fixed rate of   dollars ($) per hour. For the purposes of this Agreement, the Principals are:

      Employees' time computed at a multiple of   (   ) times the employees' Direct Personnel Expense as defined in Article 4.

   c. *FOR THE ENGINEER'S REIMBURSABLE EXPENSES,* amounts expended as defined in Article 5 when authorized in advance by the Architect.

   d. *THE TIMES AND FURTHER CONDITIONS OF PAYMENT* shall be as described in Article 6.

III. *THE ARCHITECT* shall be the Project Coordinator. The relationship of the Engineer to the Architect shall be that of an independent contractor.

TERMS AND CONDITIONS OF AGREEMENT
BETWEEN ARCHITECT AND ENGINEER

ARTICLE 1

ENGINEER'S SERVICES

1.1   General

The Engineer shall collaborate with the Architect for This Part of the Project and shall be bound to perform the services undertaken hereunder for the Architect in the same manner and to the same extent that the Architect is bound by the Prime Agreement to perform such services for the Owner so that his duties and responsibilities with respect to This Part of the Project shall be co-extensive with those of the Architect to the Owner under the Prime Agreement. Except as set forth herein, the Engineer shall not have any duties or responsibilities for any other part of the Project. The Engineer will perform his work in character, sequence and timing so that it will be coordinated with that of the Architect and other consultants for the Project. The Engineer agrees to a mutual exchange of Drawings with the Architect and other consultants in accordance with a schedule provided by the Architect.

1.2   Basic Services

The Engineer's Basic Services consist of the five phases described below.

Schematic Design Phase

1.2.1   The Engineer shall consult with the Architect to ascertain the requirements of the Project and shall confirm such requirements to the Architect.

1.2.2   The Engineer shall make recommendations regarding basic systems, attend necessary conferences, prepare necessary analyses and be available for general consultation. When necessary the Engineer shall consult with public agencies and other organizations concerning utility services and requirements.

1.2.3   The Engineer shall prepare and submit to the Architect a Statement of the Probable Construction Cost of This Part of the Project based on current area, volume or other unit costs.

1.2.4   The Engineer shall recommend to the Architect the obtaining of such investigations, surveys, tests and analyses as may be necessary for the proper execution of the Engineer's work.

Design Development Phase

1.2.5   The Engineer shall prepare from the approved Schematic Design Studies, for approval by the Architect, the Design Development Documents consisting of drawings and other documents to fix and describe This Part of the Project.

1.2.6   The Engineer shall submit to the Architect a further Statement of Probable Construction Cost of This Part of the Project.

Construction Documents Phase

1.2.7   The Engineer shall prepare from the approved Design Development Documents, for approval by the Architect, Working Drawings and Specifications setting forth in detail the requirements for the construction of This Part of the Project and shall deliver the original Drawings and Specifications to the Architect. The original Drawings and Specifications shall be in such form as the Architect may reasonably require.

1.2.8 The Engineer shall advise the Architect of any adjustments to previous Statements of Probable Construction Cost of This Part of the Project indicated by changes in requirements or general market conditions.

1.2.9 The Engineer shall assist the Architect in filing the required documents with respect to This Part of the Project for the approval of governmental authorities having jurisdiction over the Project.

Bidding or Negotiation Phase

1.2.10 If required by the Architect, the Engineer shall assist the Architect and the Owner in obtaining bids or negotiated proposals, and in awarding and preparing construction contracts.

Construction Phase–Administration
of the Construction Contract

1.2.11 The Construction Phase will commence with the award of the Construction Contract and will terminate when final payment is made by the Owner to the Contractor.

1.2.12 The Engineer shall assist the Architect in the administration of the construction contract with respect of This Part of the Project.

1.2.13 The Engineer shall at all times have access to the Work of This Part of the Project wherever it is in preparation or progress.

1.2.14 The Engineer shall make periodic visits to the site to familiarize himself generally with the progress and quality of the Work for This Part of the Project to determine in general if such Work is proceeding in accordance with the Contract Documents. On the basis of his on-site observations as an engineer, he shall endeavor to guard the Owner and the Architect against defects and deficiencies in the Work of the Contractor. The Engineer shall not be required to make exhaustive or continuous on-site inspections to check the quality or quantity of the Work. The Engineer shall not be responsible for construction means, methods, techniques, sequences or procedures, or for safety precautions and programs, and he shall not be responsible for the Contractor's failure to carry out the Work for This Part of the Project in accordance with the Contract Documents.

1.2.15 Based on such observations at the site and on the Contractor's Applications for Payment, the Engineer shall assist the Architect in determining the amount owing to the Contractor for This Part of the Project. If requested, he shall certify such amounts to the Architect. A certification by the Engineer to the Architect of an amount owing to the Contractor shall constitute a representation by the Engineer to the Architect and through him to the Owner that, based on the Engineer's observations at the site as provided in Subparagraph 1.2.14 and the data comprising the Application for Payment, the Work for This Part of the Project has progressed to the point indicated; that to the best of his knowledge, information and belief, the quality of such Work is in accordance with the Contract Documents (subject to an evaluation of such Work as a functioning whole upon Substantial Completion, to the results of any subsequent tests called for in the Contract Documents, to minor deviations from the Contract Documents correctable prior to completion, and to any specific qualifications stated by the Engineer); and that the Contractor is entitled to payment in the amount certified.

1.2.16 The Engineer shall assist the Architect in making decisions on all claims of the Owner or Contractor relating to the execution and progress of the Work on This Part of the

Project and on all other matters or questions related thereto. The Engineer shall not be liable for the results of any interpretation or decision rendered in good faith.

1.2.17 The Engineer shall assist the Architect in determining whether the Architect shall reject Work for This Part of the Project which does not conform to the Contract Documents. The Engineer shall also assist the Architect in determining whether to require the Contractor to stop the Work with respect to This Part of the Project whenever such action may be necessary for the proper performance of the Contract. The Engineer shall not be liable to the Owner or Architect for the consequences of any recommendation made by him to the Architect in good faith with reference to stopping the Work.

1.2.18 The Engineer shall check and approve shop drawings, samples and other submissions of the Contractor with respect to This Part of the Project only for conformance with the design concept and for compliance with the information given in the Contract Documents. All comments and approvals shall be submitted to the Architect.

1.2.19 The Engineer shall assist the Architect in preparing Change Orders for This Part of the Project.

1.2.20 The Engineer shall assist the Architect in conducting inspections with respect to This Part of the Project to determine the Dates of Substantial Completion and Final Completion and in receiving written guarantees and related documents assembled by the Contractor with respect to This Part of the Project.

1.2.21 The Engineer shall not be responsible for the acts or omissions of the Contractor, or any Subcontractors, or any of the Contractor's or Subcontractors' agents or employees, or any other persons performing any of the Work.

1.3 Project Representation Beyond Basic Services

1.3.1 If more extensive representation at the site than is described under Subparagraphs 1.2.11 through 1.2.21 inclusive is required for This Part of the Project and if the Architect and Engineer agree, the Engineer shall provide one or more Full-time Project Representatives to assist the Engineer.

1.3.2 Such Full-time Project Representatives shall be selected, employed and directed by the Engineer and the Engineer shall be compensated therefor as mutually agreed between the Architect and the Engineer.

1.3.3 The duties, responsibilities and limitations of authority of such Full-time Project Representatives shall be as set forth in an exhibit appended to this Agreement.

1.3.4 Through the on-site observations by Full-time Project Representatives of the Work in progress, the Engineer shall endeavor to provide further protection for the Owner against defects in the Work for This Part of the Project, but the furnishing of such project representation shall not make the Engineer responsible for the Contractor's failure to perform the Work in accordance with the Contract Documents.

1.4 Additional Services

The following services for This Part of the Project are not covered in Paragraphs 1.2 or 1.3. If any of these Additional Services are authorized by the Architect for This Part of the Project and if the Architect is additionally compensated therefor by the Owner, they shall be paid for by the Architect as hereinbefore provided.

1.4.1 Revising Drawings, Specifications or other documents previously approved by the Owner to accomplish changes not initiated by the Engineer.

1.4.2   Preparing supporting data for Change Orders where the change in the Basic Fee resulting from the adjusted Contract Sum is not commensurate with the Engineer's services required.

1.4.3   Preparing documents for alternate bids requested by the Owner.

1.4.4   Providing Detailed Estimates of Construction Costs.

1.4.5   Providing consultation concerning replacement of any Work damaged by fire or other cause during construction, and furnishing professional services of the type set forth in Paragraph 1.2 as may be required in connection with the replacement of such Work.

1.4.6   Providing professional services made necessary by the default of the Contractor in the performance of the Construction Contract.

1.4.7   Providing contract administration and observation of construction after the Contract Time has been exceeded by more than 20% through no fault of the Engineer.

1.4.8   Furnishing the Architect a set of reproducible record prints of drawings showing significant changes in This Part of the Work made during the construction process, based on marked up prints, drawings and other data furnished by the Contractor.

1.4.9   Providing services after final payment to the Contractor.

1.4.10   Providing services as an expert witness in connection with any public hearing, arbitration proceeding, or the proceedings of a court of record.

ARTICLE 2

THE ARCHITECT'S RESPONSIBILITIES

2.1   The Architect shall provide full information regarding the requirements for This Part of the Project.

2.2   The Architect shall designate, when necessary, a representative authorized to act in his behalf with respect to This Part of the Project. The Architect or his representative shall examine documents submitted by the Engineer and shall render decisions pertaining thereto promptly, to avoid unreasonable delay in the progress of the Engineer's work.

2.3   The Architect shall furnish to the Engineer a copy of the Owner's certified survey of the site.

2.4   The Architect shall request the Owner to furnish the services of a soils engineer, when such services are deemed necessary for This Part of the Project by the Engineer, including reports, test borings, test pits and other necesary operations for determining subsoil conditions.

2.5   The services, information, surveys and reports required by Paragraphs 2.3 and 2.4 shall be furnished at the Owner's expense, and the Engineer shall be entitled to rely upon the accuracy thereof.

2.6   If the Owner has given written notice to the Architect of any fault or defect with respect to This Part of the Project or non-conformance with the Contract Documents, the Architect shall give prompt written notice thereof to the Engineer.

ARTICLE 3

CONSTRUCTION COST

3.1   The Construction Cost of the Project or the Construction Cost of This Part of the Project to be used as a basis for determining the Engineer's Basic Fee for all Work designed or specified by the Engineer, including labor, materials, equipment and furnishings, shall be determined as follows, with precedence in the order listed:

3.1.1   For completed construction, the total cost of all such Work;

3.1.2   For work not constructed, the lowest bona fide bid received from a qualified bidder for any or all of such work; or

3.1.3   For work for which bids are not received, 1) the latest Detailed Cost Estimate, or 2) the Engineer's latest Statement of Probable Construction Cost.

3.2   Construction Cost does not include the fees of the Architect, the Engineer and other consultants, the cost of the land, rights-of-way, or other costs which are the responsibility of the Owner as provided in Paragraphs 2.3 through 2.5 inclusive.

3.3   Labor furnished by the Owner for the Project shall be included in the Construction Cost at current market rates. Materials and equipment furnished by the Owner shall be included at current market prices, except that used materials and equipment shall be included as if purchased new for the Project.

3.4   Statements of Probable Construction Cost and Detailed Cost Estimates prepared by the Engineer represent his best judgment as a design professional familiar with the construction industry. It is recognized, however, that neither the Engineer nor the Architect has any control over the cost of labor, materials or equipment, over the contractors' methods of determining bid prices, or over competitive bidding or market conditions. Accordingly, the Engineer cannot and does not guarantee that bids will not vary from any Statement of Probable Construction Cost or other cost estimate prepared by him.

3.5   When a fixed limit of Construction Cost is established as a condition of the Prime Agreement, the Architect may establish a fixed limit of Construction Cost for This Part of the Project which shall include a bidding contingency of ten per cent unless another amount is agreed upon in writing. When such a fixed limit is established, the Engineer shall be permitted to determine what materials, equipment, component systems and types of construction are to be included in the Contract Documents with respect to This Part of the Project, and to make reasonable adjustments in the scope of This Part of the Project to bring it within the fixed limit. If required, the Engineer shall assist the Architect in including in the Contract Documents alternate bids to adjust the Construction Cost to the fixed limit.

3.5.1   If the lowest bona fide bid, the Detailed Cost Estimate or the Statement of Probable Construction Cost exceeds such fixed limit of Construction Cost for This Part of the Project (including the bidding contingency), the Architect may require the Engineer, without additional charge, to modify the Drawings and Specifications for This Part of the Project as necessary to bring the Construction Cost thereof within such fixed limit for This Part of the Project. The providing of this service shall be the limit of the Engineer's responsibility in this regard, and having done so, he shall be entitled to his fees in the same proportion that the Architect receives payment of his fee from the Owner for This Part of the Project.

ARTICLE 4

DIRECT PERSONNEL EXPENSE

4.1   Direct Personnel Expense of employees engaged on This Part of the Project by the Engineer includes engineers, designers, job captains, draftsmen, specification writers and typists, in consultation, research and design, in producing Drawings, Specifications and other documents pertaining to This Part of the Project, and in services during construction at the site.

4.2   Direct Personnel Expense includes cost of salaries and of mandatory and customary benefits such as statutory employee benefits, insurance, sick leave, holidays and vacations, pensions and similar benefits.

ARTICLE 5

REIMBURSABLE EXPENSES

5.1   Reimbursable Expenses are in addition to the Fees for Basic and Additional Services and include actual expenditures made by the Engineer, his employees, or his consultants in the interest of This Part of the Project for the incidental expenses listed in the following Sub-paragraphs:

5.1.1   Expense of transportation and living when traveling in connection with This Part of the Project, long distance calls and telegrams, and fees paid for securing approval of authorities having jurisdiction over the Project.

5.1.2   If authorized in advance by the Architect, the expense of overtime work requiring higher than regular rates.

ARTICLE 6

PAYMENTS TO THE ENGINEER

6.1   Upon receipt of payment by the Architect from the Owner, the Architect shall pay the Engineer for services rendered for This Part of the Project in accordance with this Article 6.

6.2   Payments for Basic Services shall be made monthly in proportion to services performed so that the compensation at the completion of each phase of the Work shall equal the following percentages of the Basic Fee:

Schematic Design Phase	15%
Design Development Phase	35%
Construction Documents Phase	75%
Bidding and Negotiation Phase	80%
Construction Phase	100%

6.3   Payments for Additional Services of the Engineer as defined in Paragraph 1.4, and for Reimbursable Expenses as defined in Article 5, shall be made monthly upon presentation of the Engineer's statement of services rendered.

6.4   No deductions shall be made from the Engineer's compensation on account of penalty, liquidated damages, or other sums withheld from payments to contractors.

6.5   If the Project is suspended for more than three months or abandoned in whole or in part, the Engineer shall be paid his compensation for services performed prior to receipt of written notice from the Architect of such suspension or abandonment, together with Reimbursable Expenses then due and all terminal expenses resulting from such suspension or abandonment.

ARTICLE 7

ENGINEER'S RECORDS

7.1   Records of the Engineer's Direct Personnel and Reimbursable Expenses pertaining to This Part of the Project shall be kept on a generally recognized accounting basis and shall be available to the Architect or his authorized representative at mutually convenient times.

7.2   The Engineer shall maintain his design calculations on file in legible form and available to the Architect.

ARTICLE 8

TERMINATION OF AGREMENT

8.1   This Agreement is terminated if and when the Prime Agreement is terminated. The Architect shall promptly notify the Engineer of such termination.

8.2   This Agreement may be terminated by either party upon seven days' written notice should the other party fail substantially to perform in accordance with its terms through no fault of the other. In the event of termination due to the fault of others than the Engineer, the Engineer shall be paid his compensation for services performed to termination date, including Reimbursable Expenses then due and all terminal expenses, contingent upon comparable adjustment by the Owner of the Architect's compensation.

ARTICLE 9

INSURANCE

9.1   The Architect and the Engineer shall each effect and maintain insurance to protect himself from claims under workmen's compensation acts; claims for damages because of bodily injury including personal injury, sickness or disease, or death of any of his employees or of any person other than his employees; and from claims for damages because of injury to or destruction of tangible property including loss of use resulting therefrom; and from claims arising out of the performance of professional services causes by any errors, omissions or negligent acts for which he is legally liable.

ARTICLE 10

SUCCESSORS AND ASSIGNS

The Architect and the Engineer each binds himself, his partners, successors, assigns and legal representatives to the other party to this Agreement and to the partners, successors, assigns and legal representatives of such other party with respect to all covenants of this Agreement. Neither the Architect nor the Engineer shall assign, sublet or transfer his interest in this Agreement without the written consent of the other.

ARTICLE 11

ARBITRATION

11.1   All claims, disputes and other matters in question arising out of, or relating to, this Agreement, or the breach thereof, shall be decided by arbitration in accordance with the Construction Industry Arbitration Rules of the American Arbitration Association then obtaining unless the parties mutually agree otherwise. This agreement so to arbitrate shall be specifically enforceable under the prevailing arbitration law.

11.2   Notice of the demand for arbitration shall be filed in writing with the other party to this Agreement and with the American Arbitration Association. The demand shall be made within a reasonable time after the claim, dispute or other matter in question has arisen. In no event shall the demand for arbitration be made after institution of legal or equitable proceedings based on such claim, dispute or other matter in question would be barred by the applicable statute of limitations.

11.3   The award rendered by the arbitrators shall be final, and judgment may be entered upon it in accordance with applicable law in any court having jurisdiction thereof.

ARTICLE 12

EXTENT OF AGREEMENT

This Agreement represents the entire and integrated agreement between the Architect and the Engineer and supersedes all prior negotiations, representations or agreements, either written or oral. This Agreement may be amended only by written instrument signed by both Architect and Engineer.

ARTICLE 13

APPLICABLE LAW

Unless otherwise specified, this Agreement shall be governed by the law of the principal place of business of the Architect.

This Agreement executed the day and year first written above.

ARCHITECT   **ENGINEER**

Architect's Registration No.   Engineer's Registration No.

CONSULTING ENGINEERS COUNCIL
A STANDARD FORM OF
AGREEMENT BETWEEN ARCHITECT AND ENGINEER

Fee is Per Cent of Construction Cost

THIS AGREEMENT made as of the_____ day of _____
19 _____ by and between _____

hereinafter called the ARCHITECT, and _____
_____hereinafter called the ENGINEER,
WITNESSETH, that whereas the ARCHITECT has made an agreement with _____

hereinafter called the OWNER, to provide professional services in connection with _____

_____hereinafter called the PROJECT.
NOW, THEREFORE, the ARCHITECT and ENGINEER for the considerations hereinafter set forth agree as follows:

1. THE ENGINEER AGREES to perform the following Engineering services for the_____

_____work of the PROJECT, hereinafter called THIS PART OF THE PROJECT:

a. General: The Engineer shall serve as the Architect's professional associate and he shall perform for the Architect all the Engineering services for This Part of the Project which the Architect is required to furnish under his agreement with the Owner, a copy of the conditions of which is attached hereto and is made a part of this Agreement. The Engineer shall so perform his work as to coordinate the same with that of the Architect and other Engineers for the Project arranging the character, order and schedule thereof for that purpose, and he shall attend the necessary conferences and provide consultation and advice for This Part of the Project.

(1) Engineer's Design Calculations: The Engineer agrees to maintain his design calculatons on file in legible form and available to the Architect.

(2) Copyright or Patent Infringement: The Engineer shall defend actions or claims charging infringement of any copyright or patent by reason of the use or adoption of any designs, drawings or specifications supplied by him, and he shall hold harmless the Owner and Architect from loss or damage resulting therefrom providing however, that the Owner or Architect, within five (5) days after receipt of any notice of infringement or of summons in any action therefor, shall have forwarded the same to the Engineer in writing.

(3) Insurance: The Engineer shall secure and maintain such insurance as will protect him from claims under the Workmen's Compensation Acts and from claims for bodily injury, death, or property damage which may arise from the performance of his services under this Agreement.

b. Basic Services of the Engineer:

(1) Diagrammatics: The Engineer shall prepare schematic drawings and layouts and a construction cost estimate based upon the diagrammatics.

---

(2) Preliminaries: The Engineer shall establish approximate space requirements for equipment and appurtenances, and shall prepare preliminary drawings, outline specifications and a construction cost estimate based upon the preliminaries.

(3) Contract Documents: The Engineer shall prepare from the approved preliminaries, working drawings and specifications completely describing the material and workmanship required for This Part of the Project, and he shall adjust the preliminary construction cost estimate to include changes in the scope of the Project, the Owner's requirements and market conditions.

(4) Receipt of Proposals: The Engineer shall assist the Architect in analyzing Proposals received for the construction of This Part of the Project.

(5) During Construction: The Engineer shall provide general supervision of construction for This Part of the Project to check the Contractor's work for general compliance with the Contract Documents and shall endeavor to protect the Owner against defects and deficiencies in the work of the Contractor, but he does not guarantee the Contractor's performance. The Engineer's general supervision shall not include furnishing a full time resident Engineer, but shall include the following services:

(a) Additional Instructions: The Engineer shall assist the Architect in issuing such additional instructions as may be necessary to interpret the drawings and specifications or to illustrate required changes in This Part of the Project.

(b) Contractor's Submittals: The Engineer shall check shop drawings, samples, and other data submitted by the Contractor for compliance with the drawings and specifications.

(c) Contractor's Requests for Payment: The Engineer shall assist in approving the Contractor's requests for payment for work performed under This Part of the Project.

(d) Visits to Site: The Engineer shall make periodic inspections at the site to check the Contractor's work for general compliance with the Contract Documents and to determine the extent of work completed for checking of Contractor's requests for payment.

(e) Special Performance Tests: The Engineer shall witness and fully report the results of all special performance tests required for This Part of the Project.

(f) Final Acceptance: The Engineer shall prepare completion lists when 90% completion of This Part of the Project is claimed by the Contractor and again when 100% completion is claimed. Following Contractor's completion of the items outlined in the completion lists, the Engineer shall certify compliance with the drawings and specifications, following a final visit to the Project.

(g) Instructions to the Owner: The Engineer shall arrange for detailed instruction by the Contractor and manufacturers' representatives of the Owner or his delegated representative in the proper operation and maintenance of the equipment furnished and installed under This Part of the Project.

c. Extra Services of the Engineer shall include the following items in the interest of This Part of the Project when authorized in writing by the Architect;

(1) Contract Documents: Revisions to drawings or specifications previously approved and preparation of drawings and specifications for alternate proposals and change orders.

(2) During Construction: Resident supervision of construction of This Part of the Project; supervising the replacement of all or such parts of This Part of the Project as may be

damaged by fire or other cause during the construction; assisting the Architect in arranging for continuation of the work should the Contractor default for any reason; providing supervision of construction over an extended period should the construction contract time be exceeded by more than 25% not occasioned by fault of the Engineer, and preparing record drawings showing construction changes in the work and showing equipment, mechanical or electrical service lines and outlets, as finally located.

(3) Inspection Prior to Expiration of the Guaranty Period of This Part of the Project and preparation of a written report listing discrepancies between guaranties and performance.

d. Engineer's Reimbursable Services shall include the following items in the interest of This Part of the Project when authorized in writing by the Architect: Transportation and subsistence of principals and employees on special trips to the Project or to other locations as requested by the Architect or the Owner; long distance telephone and telegraph calls as required to expedite the Contractor's work under This Part of the Project.

2. THE ARCHITECT AGREES to provide the Engineer with complete information concerning the requirements of This Part of the Project and to perform the following services:

a. Site Survey: The Architect shall furnish to the Engineer a topographic and boundary survey of the site including all data pertinent to This Part of the Project, and in addition all required information concerning service and utility lines, both public and private.

b. Tests and Reports: The Architect shall obtain authorization for structural, chemical, mechanical, soil mechanics, percolation and other tests and reports to be prepared by others, in accordance with recommendations of the Engineer, and shall furnish to the Engineer all the resulting data required for his work, with such promptness as to avoid undue delay in the work of the Engineer.

c. Consideration of the Engineer's Work: The Architect shall give thorough consideration to all reports, sketches, estimates, drawings, specificatons, proposals and other documents presented by the Engineer, and shall inform the Engineer of his decision within a reasonable time so as not to delay the work of the Engineer.

d. Reproduction: The Architect shall pay the cost of reproducing all drawings and specifications, both preliminary and final, that are required for the Owner, governmental approving agencies, and for bidding and construction purposes.

e. Standards: The Architect shall furnish the Engineer with a copy of any design and construction standards he shall require the Engineer to follow in the preparation of drawings and specifications for This Part of the Project.

3. THE ARCHITECT'S PAYMENTS TO THE ENGINEER:

a. General:

(1) Definition of Construction Cost of This Part of the Project as herein referred to means the total cost of all work designed or specified by the Engineer but does not include any payments to the Architect, the Engineer, or other consultants.

(2) Payments Withheld From Contractors: No deduction shall be made from the Engineer's compensation on account of penalty, liquidated damages, or other sums withheld from payments to Contractors.

(3) Abandoned or Suspended Work: If any work performed by the Engineer is abandoned or suspended in whole or in part, the Engineer shall be paid for the services performed prior to receipt of written notice from the Architect of such abandonment or suspension, together with any terminal expense resulting therefrom.

(4) Progress Payments: Once each _____ , the Engineer shall present to the Architect a statement for professional services performed under Articles 1.b, 1.c and 1.d of this Agreement. The amount of the statement shall be in proportion to the services performed during the period. Payment shall be due the Engineer when the statement is rendered.

b. Payments for Basic Services of the Engineer: The Architect shall pay the Engineer for the basic services described in Article 1.b _____ per cent (_____ %) of the Construction Cost of This Part of the Project, hereinafter referred to as the Basic Rate, with progress payments as herein provided. At the completion of each phase of the work, progress payments shall total the following percentages of the Construction Cost of This Part of the Project:

    (1) Diagrammatics: _____ %, based upon the Diagrammatic Cost Estimte.
    (2) Preliminaries: _____ %, based upon the Preliminary Cost Estimate.
    (3) Contract Documents: _____ %, based upon the final Cost Estimate.
    (4) Receipt of Proposals: _____ %, based upon the lowest bona fide Proposal.
    (5) During Construction: _____ %, based upon the total Construction Cost
                         of This Part

c. Payments for Extra Services of the Engineer: For Extra Services defined in Article 1.c., the Architect agrees to pay the Engineer principals' time computed at $ _____ per hour plus (_____) times the direct personnel expense for employees, including charges for Engineers, designers, draftsmen, specification writers, typists, and inspectors.

d. Payments for Engineer's Reimbursable Services: The Engineer shall be reimbursed at cost for the reimbursable services outlined under Article 1.d.

4. THE ARCHITECT AND ENGINEER FURTHER AGREE to the following conditions:

a. Mutual Assistance: Each party shall render assistance to the other from any and all actions or claims arising out of such party's errors or omissions.

b. Termination: In the event of termination of the Agreement between the Owner and Architect, this Agreement shall be terminated by Architect sending a notice in writing to Engineer of such termination of the Owner's Agreement. This Agreement may be terminated by either party by seven (7) days' written notice in the event of substantial failure to perform in accordance with the terms hereof by the one party through no fault of the other party. If terminated due to the fault of others than the Engineer, the Engineer shall be paid for services performed to the date of termination, including reimbursements then due, plus terminal expense.

c. Arbitration: Arbitration of all questions in disupte under this Agreement shall be at the choice of either party and shall be in accordance with the rules of the American Arbitration Association. This Agreement shall be specifically enforceable under the prevailing arbitration law and judgment upon the award rendered may be entered in the court of the forum, state or federal, having jurisdiction. The decision of the arbitrators shall be a condition precedent to the right of any legal action.

d. Ownership of Documents: The completed tracings and master specification sheets shall _____
the property of the_____.
(1) Re-use of Drawings and Specifications: The Engineer's tracings or master specification sheets or reproductions of them in whole or in part shall not be used on additions to this Project or on any other project, except upon written agreement with the Engineer.

5. SUCCESSORS AND ASSIGNS: This agreement and all of the covenants hereof shall inure to the benefit of and be binding upon the Architect and the Engineer respectively and his partners, successors, assigns and legal representatives. Neither the Architect nor the Engineer shall have the right to assign, transfer or sublet his interests or obligations hereunder without written consent of the other party.

6. SPECIAL PROVISIONS: The Architect and Engineer mutually agree that this Agreement shall be subject to the following Special Provisions, which shall supersede other conflicting provisions of this Agreement:

IN WITNESS WHEREOF the parties hereto have made and executed this Agreement, the day and year first above written.

ARCHITECT: ENGINEER:

_____   _____

_____   _____

_____   _____

_____   _____

_____   _____

_____   _____

CCAIA Doc. No. 105-65,
© 1965 by California Council, The American Institute of Architects

The California Form of Agreement for Consultant Services

Prepared by California Council, The American Institute of Architects in cooperation with Consulting Engineers Association of California, Structural Engineers Association of California, California Council of Civil Engineers and Land Surveyors, and California Council of Landscape Architects, for use between Consultant and Architect.

PROJECT _____

The Consultant and the Architect agree as follows:

1. CONSULTANT'S SERVICES: The Consultant's professional service to the architect shall consist of: _____

_____

(herein called the "Consultant's services") as follows:

a. Collaboration with the Architect, who shall be the Project coordinator, with respect to the Consultant's services and performance of the services undertaken hereunder for the Architect in the same manner and to the same extent that the Architect is bound by his contract with his Client to perform such services for his Client so that the Consultant's duties and responsibilities with respect to the Consultant's services shall be co-extensive with those of the Architect under his contract with his Client, a copy of which (except as to amount of compensation) is attached hereto and made a part hereof by reference. The Consultant shall perform his services in character, sequence and timing so that they will be coordinated with the services of the Architect and other Consultants for the project and so that his services shall conform to the original or revised schedule and budget for the Project as provided by the Architect. The relationship of the Consultant to the Architet shall be that of an independently practicing professional performing services for the Architect in accordance with this agreement.

b. Scheme and Program Development.

c. Preliminary Studies.

d. Working Drawings; Specifications and Calculations

e. Administration of the construction work pertaining to the Consultant's services.

2. COMPENSATION:

a. Basic compensation shall be _____

and shall be paid as follows: _____

b. If any work designed by the Consultant is abandoned or suspended, or if this agreement is cancelled, the Consultant shall be paid an equitable amount in accordance with paragraph 2a.

3. ADDITIONAL SERVICES:

a. The Consultant shall be reimbursed for unusual expenses as follows: _____

and shall be paid as follows:

4. PROJECT DATA: The Consultant shall obtain from the appropriate sources all data and information necessary for the proper and complete execution of the Consultant's

services, except that the Architect shall provide basic programming, scope, and budget limitations of the project.

5. ADMINISTRATION: The Consultant shall provide general administration of the construction phase to ascertain whether it is being executed in conformity with drawings, specifications or directions pertaining to the Consultant's services, but the Consultant is not responsible for the performance of contractors or for their errors or omissions. If an approved project inspector is employed on the project, the Consultant shall provide technical direction within the scope of the work, based on the Consultant's services.

6. PRINTS AND INFORMATION: The Consultant shall furnish at his expense to the Architect and other Consultants all progress prints and information required for the Consultant's services. The Architect shall furnish at his expense information and progress prints of his work required for the execution of the Consultant's services. Prints required by reviewing agencies shall be paid for by ⸺⸺⸺⸺⸺

Final prints and specifications furnished to the Consultant shall be paid for by ⸺⸺.
Documents required for bidding and construction shall be paid for by ⸺⸺⸺.

7. SUCCESSORS AND ASSIGNMENTS: The Consultant and the Architect each binds himself, his partners, successors, legal representatives, and assigns to the other party to this agreement, and to the partners, successors, legal representatives and assigns of such other party in respect of all covenants of this agreement. Neither the Consultant nor the Architect shall assign or transfer his interest in this agreement without the other's written consent.

8. INSURANCE: Each party to this agreement shall carry and maintain public liability, property damage, and workmen's compensation insurance and in addition insurance in an amount of at least $ ⸺⸺⸺⸺to protect him from claims arising out of the performance of his professional services caused by the acts, errors, or omissions of each party or his employees, or by others for whom he is legally responsible.

9. ARBITRATION: All questions in dispute under this agreement shall be submitted to arbitration when practical in accordance with the provisions, then obtaining, of the American Arbitration Association. The prevailing party shall be entitled to reasonable attorneys' fees to be fixed by the Arbitrator; or in the event there are judicial proceedings instead of arbitration reasonable attorneys' fees shall be fixed by the court.

10. SCOPE OF AGREEMENT: This is the entire agreement between the parties and there are no conditions, agreements or representations between the parties except as expressed herein.

11. CANCELLATION: This agreement may be cancelled by either party by written notice mailed to the other party at his place of business.

12. ADDITIONAL PROVISIONS: ⸺⸺⸺⸺⸺⸺⸺⸺

⸺⸺⸺⸺⸺⸺⸺⸺⸺⸺⸺⸺

The above is mutually agreed to this ⸺⸺ day of ⸺⸺⸺,196 ⸺.

By ⸺⸺⸺⸺⸺⸺   ⸺⸺⸺⸺⸺⸺
⸺⸺⸺⸺⸺⸺   ⸺⸺⸺⸺⸺⸺
Architect   Consultant

INSTRUCTION SHEET

For use with CCAIA Doc. No. 105-65

*This instruction sheet is not a part of the agreement
and does not vary or modify it in any way.*

The California Form of Agreement for Consultant's Services has been designed to serve as an agreement for all types of consulting services. It is intended to be used in conjunction with CCAIA Architect-Client Documents No. 102 and 103. If a different Architect-Client agreement is used, care should be taken to make sure that the scope of the work and the terminology agree with that used in the Consulting Services form.

If your special needs cannot be written within the blank spaces provided, or if more elaboration is required, consideration should be given to either the National AIA forms or special agreement to meet your needs.

For assistance in the preparation of special forms, refer to the pamphlet prepared by the AIA in 1961, *"Concerning Some Legal Responsibilities in the Practice of Architecture and Engineering and Some Recent Changes in the Contract Documents."*

Caption

The project or work should be described in this space.

Paragraph 1.

Herein insert the nature of the Consultant's Services, i.e., structural engineering, mechanical engineering, etc.

Paragraph 1a.

Attach a copy of the Architect-Client agreement with compensation provision optionally removed.

Paragraph 2a.

Describe here the basis of the fee, such as percentage basis or lump sum. For methods of payment insert applicable terms, i.e.: "at the times and in accordance with the percentages of compensation due the Architect from his Client"; "upon monthly billing"; or "upon completion of each phase, a sum equal to ——————", or other basis.

Paragraph 3a.

Insert herein "Pursuant to the provisions of the Architect-Client agreement," or enter those reimbursable expenses, such as travel expense, long distance telephone calls and telegrams, and other expenses to be charged to the Architect if such is the agreement with the Consultant, including time and method of payments for same.

Paragraph 3b.

Insert here: "regularly"; "monthly"; "from time to time"; or "as the work is performed or expenses incurred."

Paragraph 4.

Insert here additional items to be provided by either party, i.e., surveys, soils reports, etc.

Paragraph 6.

Insert herein: "owner"; "architect" or "consultant" or as mutually agreed.

Paragraph 12.

List or attach an addendum of any additional provisions, for example: as-built drawings; budget limitations; definition of changes in the work, etc.

Chapter 4

Conditions of the Construction Contract

Identification of the Contract Documents

The complex nature of building construction calls for a relatively lengthy contract if the rights and obligations of the owner and contractor are to be satisfied with sufficient explicitness to make it an effective operational instrument. As a consequence, construction contracts usually consist of several documents which are complementary and, taken together, represent an integrated agreement.

Although the terms *agreement* and *contract* are commonly used synonymously, custom within the building industry has developed an additional connotation with respect to agreement, using it to denote the basic constituent document that welds all of the Contract Documents into a single integrated contract. Although practice varies somewhat, the essential items that constitute the Contract Documents may be listed as follows:

1. The Agreement
2. The Conditions of the Contract
     General Conditions
     Supplementary Conditions
3. The Drawings
4. The Specifications
5. Addenda (amendments to the Contract Documents issued prior to execution of the agreement)[1]

[1]To help clarify discussion, any changes in the contract or supplementary agreements made subsequent to its execution are often referred to as Modifications, but this usage is by no means universal.

The Agreement is the basic document. It incorporates all of the other documents by reference, thereby constituting them a single, integrated contract. In addition to identifying the parties concerned, the contract sum is stated therein, together with any special provisions having to do with time of completion and method of payment.

The General Conditions define the rights, responsibilities, and relationships of the parties concerned. The group of standardized provisions issued by the American Institute of Architects as AIA Document A201, General Conditions of the Contract for Construction, represent items identified over years of construction industry experience that must be handled by prior agreement if disputes are to be avoided. In federal government construction contracts, analogous provisions are given under the title of *General Provisions.*

The Supplementary Conditions provide a mechanism for modifying or extending provisions of standardized General Conditions to fit the special requirements of specific projects.

The Drawings, commonly called Working Drawings, portray graphically the work to be constructed with respect to dimensions, location, arrangement of components, materials, mechanical and electrical systems, and siting.

The Specifications provide technical information concerning the building materials, components, systems, and equipment indicated on the Drawings with respect to quality, performance characteristics, and results to be achieved by application of construction methods.

The foregoing brief description of the Contract Documents is given to provide an overview of their general character and content. Each of them (except the Working Drawings) will be considered more fully with the Conditions of the Contract in this chapter and the others in subsequent chapters.

Conditions of the Contract as a Separate Document

Although there is no reason why the contractual stipulations and covenants contained in the Agreement and the Conditions of the Contract cannot be consolidated into one document (a common practice in contracts for heavy construction), the bulk of all private building construction in the United States is accomplished under the two-document pattern. This permits free access to the General Conditions and Supplementary Conditions by all concerned on the project although restricting access to financial and other special arrangments contained in the Agreement. It has long been customary to refer to the General Conditions and Supplementary Conditions at the beginning of each section of the Specifications, pointing out that the provisions of these two documents apply to the work described under that particular section. Also, the General

Conditions will usually apply to each prime contract when the project is constructed under the separate contracts system, although the several Agreements will be different.

The discussion in the remainder of this chapter is focused on lump-sum construction contracts and is equally applicable whether award is made by competitive bidding or negotiation. For the most part, the provisions apply also to cost-plus-fee contracts, but some modification will be required in this connection as indicated in Chapter 9. Although the discussion is general and is independent of any standard form, two forms are included at the end of this chapter.[2] It is recommended that the reader study them and compare the provisions discussed here with their corresponding provisions. Here, as elsewhere in this book, direct reference to article numbers of standard forms is not made, but brief examination of the documents will clarify their pattern of organization and facilitate cross reference.

The General Conditions, regardless of the exact form in which they occur, are important to all segments of the building industry. Although they are specifically a part of the owner-builder contract they define, either directly or indirectly, the interrelationships among the several parties concerned with the construction of a project.

Articles of the General Conditions

Although there is considerable flexibility in the topics contained in the General Conditions and in their arrangement, articles most frequently used in current practice may be classified under the following headings:

Definition and Status of the Contract Documents
Administrative Functions of the Architect
Rights and Duties of the Owner
Rights and Duties of the Contractor
Subcontracting Procedures
Performance and Payment Bonds
Royalties and Patents
Arbitration of Claims and Disputes
Contract Time, Delays, and Extensions
Payments to the Contractor
Safety Precautions and Insurance
Changes in the Work
Uncovering and Correction of Work
Termination of the Contract
Miscellaneous Provisions

[2] These are AIA Document A201 General Conditions of the Contract for Construction; and U.S. Government, General Services Administration Form 23-A, General Provisions (Construction Contract).

Many of these topics may require several separate articles or clauses in the General Conditions in order to cover adequately the detailed provisions intended. The commentary on them that follows relates directly to construction under a single general contract. Modifications and additions to the General Conditions when used under the separate contracts system are discussed later in this chapter.

Definition and Status of the Contract Documents

It is essential that the Contract Documents be identified and that the relationship between the several documents be stated clearly. For definition purposes a listing similar to that given in the first paragraph of this chapter is normally sufficient if supplemented by specifying the nature of Addenda and Modifications. A generally accepted definition of an Addendum is a revision, supplement, or amendment to the Drawings and/or Specifications issued subsequent to their distribution to prospective bidders but *before* execution of the contract. A Modification is a written amendment made *after* execution of the contract and may consist of a supplemental agreement between owner and contractor, a change order, or a written interpretation of the Drawings and Specifications by the architect-engineer.

The Contract should be defined as consisting of the Contract Documents and it should be stipulated that as so defined it represents the entire agreement made between owner and contractor, superseding all prior negotiations or representations, including the bidding documents. All articles of the General Conditions should be confined to procedures operative after execution of the contract, and the words "bid" or "bidder" should not appear in any of the documents making up the contract.[3]

Concerning the status of the individual documents, it is necessary to emphasize that the Contract Documents are complementary, that is, there is no order of precedence. Requirements called for in one of the documents are as binding as if required by all. There is a time-honored but hopefully vanishing custom that seeks to establish the precedence of the Specifications over the Drawings (or vice versa) in order to resolve conflicts that may appear in these documents. Such practice, however, is not compatible with the complexities of modern building and inconsistencies between the documents should be resolved on the merits of the individual situation as it occurs.

Administrative Functions of the Architect

As pointed out under the discussion of construction phase services in Chapter 2, the architect-engineer is not a party to the construction contract between owner

[3]For matters pertaining to instructions to bidders and other bidding documents, see Chapter 7, Bidding and Award Procedures.

and contractor. Nevertheless, in the interest of avoiding misunderstandings, the General Conditions of the construction contract should spell out clearly the functions that the architect is to perform in the administration of that contract. It is important that a statement be made identifying the "architect" or "architect-engineer"[4] as the person or organization cited as such in the Agreement. Equally important is a clause stipulating that nothing contained in the Contract Documents shall create any contractual relationship between the architect and the contractor. One additional broad provision should state that the architect will provide general administration of the contract and that he will act as the owner's representative during construction, with the owner issuing all instructions to the contractor through the architect.

Several individual items relating to the duties and authority of the architect in exercising his contract administration functions, and which require specific contract clauses, are included in the following discussion. It is obviously of prime importance that the wording of the General Conditions in this regard be consistent with the provisions of the owner-architect contract.

ON-SITE OBSERVATIONS

There should be a clear statement concerning the purpose of the architect's periodic visits to the site. As noted in Chapter 2, these visits were formerly performed under the heading of Supervision of Construction. However, in view of the apparent reluctance of the courts to construe this term in accordance with traditional building industry usage, its continued employment is of doubtful wisdom.

The article of the General Conditions describing the purpose of the periodic visits should be worded exactly the same as the corresponding article in the contract for architectural and engineering services; thus there is assurance that the construction contractor will have full knowledge of the architect's responsibility to the owner with regard to progress and quality of the work. It should be made clear that exhaustive and continuous on-site inspections to check quality of the work are not contemplated and that the architect will not be responsible for construction means, methods, techniques, or sequences. Although the architect will endeavor to guard the owner against defects in the work, he will not assume responsibility for its execution in accordance with the Contract Documents. It is the construction contractor who has made a contract with the owner to so construct the project.

CERTIFICATES FOR PAYMENT

Although the subject of payments to the contractor is usually covered elsewhere in the General Conditions and will be discussed more fully below, a statement in

---

[4]Whichever term is to be used in a particular situation. As elsewhere in this book, the designations *architect* and *architect-engineer* are used synonymously in this discussion.

this connection should appear under the general articles describing the administrative functions of the architect. This statement should make clear that on the basis of the on-site observations and the contractor's application for progress payments, the architect will determine the amounts due the contractor and will issue certificates for payment.

INTERPRETATION OF CONTRACT DOCUMENTS

In contracts for private construction it is customary to designate the architect-engineer as the interpreter of the Contract Documents and the judge of the performance of both owner and contractor. This, of course, places the architect in the anomalous position of acting as an impartial referee while also serving as the authorized representative of one party to the contract. The long record of satisfactory operations under such an arrangment attests to the professional integrity of the architects and engineers involved.

In connection with this provision it should also be stipulated that interpretations and decisions thereunder must be consistent with the intent of the Contract Documents and that the architect-engineer will not show partiality to either party in arriving at his conclusions.

STATUS OF DECISIONS

It is common practice to provide that the architect's decisions on claims or disputes arising from his interpretations of the Contract Documents shall be subject to arbitration. However, to facilitate the reaching of agreement in such matters short of recourse to arbitration proceedings, the parties to the contract usually agree that a demand for arbitration may not be made before the time the architect has rendered his decision in writing or until a specified number of days have elapsed after presenting the dispute to the architect without his rendering his decision. Further discussion of arbitration procedures is presented in a subsequent article.

REFERENCED ITEMS

Since procedures concerning the handling of shop drawings, change orders, and inspections before completion are usually treated in other articles of the General Conditions, the necessity for referring to them under the general heading of architect's administrative functions is only to clarify his authority and duty to participate. This may be accomplished by clauses such as "The Architect will prepare change orders in accordance with Article . . ," etc. Procedures for handling shop drawings are discussed in the article on responsibilities of the contractor; those relating to change orders, under changes in the work; and inspections for the purpose of establishing dates of substantial completion and final completion are discussed under payments and completion.

Rights and Duties of the Owner

The principal purpose of this article is to define the owner, enumerate the information and services required of him, and refer to other articles of the General Conditions relating to his responsibilities in connection with making payments and arranging for certain types of insurance coverage during the life of the contract. The matter of definition should be handled by reference to the person or organization identified as owner in the Agreement, and it should also be stated that the term encompasses the authorized representative of the owner.

Although the owner has presumably furnished the architect with the required surveys describing physical characteristics of the site, utility locations, and legal limits for incorporation into the plans and specifications, it is advisable to restate this responsibility of the owner in the General Conditions. Closely allied to this, there should be a provision requiring the owner to secure and pay for any necessary easements.

Since the owner is called on from time to time for approvals and for special information or services under his control, a provision should be included requiring that these be furnished with reasonable promptness to avoid delaying the contractor's scheduled progress of the work. This is essential if the contractor is operating under a fixed completion date since he must have some basis for requesting an extension of time if owner-furnished information or owner approvals are delayed unreasonably.

This article should also contain the provision that the owner shall issue all instructions to the contractor through the architect. If this practice is not adhered to, centralized administration of the contract cannot be achieved. Furthermore, by issuing instructions directly to the contractor the owner may inadvertently assume some of the functions of directing and controlling the means, methods, techniques, and sequences of construction, thereby relieving the contractor from some of his legal responsibilities. There is also the hazard that the owner might become liable for negligent acts of the contractor occurring during the course of construction.

The owner may delegate some of his authority to the architect as his representative during construction. Some delegations are set forth explicitly in the General Conditions, covering such items as approval of subcontractors, approval of change orders, and temporary suspension of the work thereunder. Although many contracts for the construction of buildings do not contain special provision for suspension of the work by the owner if it becomes in his interest to do so, it may be advisable to include such provision in some contracts, especially those for extensive multi-building projects. Articles covering temporary suspension should assert the owner's right to suspend the work on notice to the contractor and should require that the contractor resume work within a specified number of days after the date fixed in a written notice from the owner to do so. Provision should be made for reimbursement to the contractor for expenses incurred in connection with the suspension and also for the contractor to abandon the work if the suspension is not lifted within a specified period.[5]

Rights and Duties of the Contractor

In establishing these articles of the General Conditions it should be borne in mind that the principal objective in letting a building construction contract is to secure the services of a person or organization skilled in methods and techniques of construction and in the management of construction operations. Consequently, the provisions under this heading should clearly establish the scope of the contractor's obligations in constructing the project in accordance with the Drawings and Specifications and should spell out procedural matters for operating purposes.

Not all of the contractor's rights and duties are enumerated under this heading because several of them are the subject of other articles of the General Conditions. However, in addition to defining the contractor as the person or organization identified as such in the Agreement, several other items should be incorporated as subarticles or clauses. These are discussed in the following paragraphs.

SUPERVISION AND DIRECTION

This provision should place full responsibility on the contractor for supervision and direction of the construction process; it should also stipulate that he is solely responsible for the sequence of operations, construction methods and techniques, and coordination of all work under the contract.

LABOR, MATERIALS, AND TAXES

Traditionally, the provision in this regard was framed along the lines of "The contractor is to furnish all materials, equipment, and labor necessary to construct. . . ." Current practice, however, favors more explicit language and usually stipulates that the contractor "shall provide and pay for all materials, labor, equipment, water, heat, utilities, tools, construction equipment, transportation, and other facilities . . ." necessary to carry out the work. Exceptions are noted with respect to any owner-furnished materials or services, and it is usual to

[5]For an example of suspension clauses see Section 23 of General Conditions of Contract for Engineering Construction, issued jointly by the American Society of Civil Engineers and the Associated General Contractors of America as a part of ASCE Form JCC-1 and AGC Standard Form 3, 1966 Ed.

provide that all materials and equipment incorporated in the project must be new. In order to assure that sales, consumer, and use taxes are considered as part of the cost of materials and equipment, there is also included a clause requiring the contractor to pay these taxes.

CASH ALLOWANCES

Since there are usually some materials or products about which full decision is not reached before the bidding period, cash allowances are stipulated to cover their estimated cost. Hardware and special finish items are often handled this way, with the specific amount of the allowance called for in the Specifications. The General Conditions, however, should carry a provision requiring the contractor to include all such allowances in his contract sum and stipulate just what costs are contemplated. Usually, cash allowances are meant to cover the net cost of products delivered at the site, with the contractor covering all additional costs including installation, overhead, and profit elsewhere in the contract sum. The Specifications should indicate the specific items of expense covered by individual cash allowances, and any necessary modifications of the clause in the General Conditions can be made in the Supplementary Conditions.

PERSONNEL AT PROJECT SITE

The contractor should be required to employ a project superintendent who will be resident at the site during the construction period and who will represent the contractor so that communications given to the superintendent will be as binding as if given to the contractor. It should be required that the superintendent be satisfactory to the architect-engineer and that he shall not be changed without his consent. There also should be a provision that makes the contractor responsible for both the acts and omissions of his employees and the subcontractors and their employees.

DOCUMENTS AT THE SITE

In order to facilitate effective administration of the contract, the contractor should be required to keep on file at the site a copy of all drawings, specifications, addenda, change orders, approved shop drawings, and field orders. These documents are to be available to the architect-engineer or his authorized representative at all times and become the property of the owner on completion of the project.

PROGRESS SCHEDULE

A schedule indicating dates for starting and completing the several operations and phases of construction should be prepared by the contractor immediately after award of contract and submitted to the architect-engineer for approval. The number of days allowed for preparation of the schedule and the character of its structure (that is, bar chart or critical path network) may be specified in the

Supplementary Conditions. This schedule must, of course, be related to other contract requirements pertaining to time of completion and should, as far as possible, be based on work classifications consistent with the schedule of values discussed in the article on payments and completion.

SHOP DRAWINGS AND SAMPLES

The definition of shop drawings should include all drawings, diagrams, performance charts, and other information prepared by the contractor, any subcontractor, or manufacturer to illustrate some element or component of the project. Samples should be defined as physical examples of materials, components, and equipment submitted by the contractor to illustrate the character, quality, and workmanship proposed to meet specified requirements.

The procedure for submittal and checking of shop drawings and samples should be spelled out in sufficient detail to define the flow of these documents and the extent of both contractor's and architect's responsibility during the process. Since a large proportion of the shop drawings normally originates with product manufacturers, suppliers, or subcontractors, the contractor should be required to review them before submission to the architect and to indicate to him any deviations from the requirements of the Contract Documents.

It should be stated clearly that the architect's approval of shop drawings and samples is only for conformance with the design concept of the project and overall compliance with the Contract Documents. His approval does not relate to their accuracy in matters of fabrication or construction techniques nor does it imply approval of changes from the requirements of the Contract Documents unless such changes have been called to the architect's attention in writing by the contractor. It should be emphasized further that the architect's approval of shop drawings and samples does not relieve the contractor from responsibility for errors and omissions.

The number of copies of shop drawings required will vary with different elements and components of the project and may be called for in the Specifications.

COMMUNICATIONS

Consistent with similar provisions in connection with the duties of architect and owner, there should be a specific statement in this article requiring that the contractor forward all communications to the owner through the architect.

INDEMNIFICATION

The growing number of lawsuits against owners, architects, and engineers arising principally from jobsite accidents has generated a tendency to incorporate indemnification and hold-harmless clauses in construction contracts. A major factor is the extent to which workmen's compensation laws limit the liability of contractors for death or injury of their employees. This situation apparently encourages instigation of "shotgun" lawsuits against owner, architect-engineer, and others involved for damages far in excess of the contractor's legal liability.

The hold-harmless clauses also seek to indemnify the owner and architect against judgments obtained by third parties in suits based on personal injury or death, property damage, and loss of use of property caused by any negligent act or omission of the contractor. The difficulty in writing equitable provisions in this connection arises from the possibility that the event in question may have been caused *in part* by a defect in the Drawings and Specifications prepared by the architect-engineer or by some action by the owner of the project. Objections by contractors usually press this point, thus alleging that includion of hold-harmless clauses intended to protect owner and architect whether or not the damage claimed is caused in part by them is an unjustifiable shift of professional responsibility.

The matter is complicated further by the questionable legality of broad indemnification clauses related to construction in certain states. There is some opinion that their use is not in the public interest and, therefore, where statute so provides, may be void and unenforceable. The situation has yet to be clarified. It should be noted, however, that AIA Document A 201 reproduced at the end of this chapter contains an indemnification provision.

CONTRACTOR'S RIGHTS

Most of the contractor's rights are stated in other articles of the General Conditions discussed here. The principal ones concern extensions of time, recourse in event of owner's failure to make payments, conditions under which he may terminate the contract, and appeals from the architect's decisions with respect to interpretation of the Contract Documents and other claims.

Subcontracting Procedures

Like the architect a subcontractor is not a party to the construction contract between owner and contractor. In fact, one of the first clauses under this article usually emphasizes that nothing contained in any of the Contract Documents shall create a contractual relationship between the owner and a subcontractor. The principal reason for treating this subject at all in the General Conditions is to provide some measure of control by the owner and architect over subcontracting operations. This is accomplished by making mandatory certain procedures the contractor is to follow in selecting subcontractors and in executing contracts with them. As noted earlier the contractor is responsible for the acts and omissions of all subcontractors. He is also obligated to pay them for their work as independent contractors to him. This article should include a clause establishing clearly that neither the owner nor the architect has any obligation under the contract to see that subcontractors are paid by the contractor, except as may be required by lien laws operative at the place of building.

CONTROL OF SUBCONTRACT AWARDS

One widely used provision for this purpose requires that the contractor submit the names of the subcontractors he proposes to use to the architect before awarding any subcontracts. This list is then canvassed by architect and owner, and if a change of any subcontractor is necessary an adjustment is made in the amount of the prime contract that reflects any difference occasioned by the change.

Current procedures of the American Institute of Architects relating to approval of subcontractors require that the successful bidder furnish his list of proposed subcontractors prior to award of the prime contract. If the owner refuses to accept a firm on the list, the successful bidder may withdraw his bid or submit an acceptable substitute contractor either with or without an increase in his bid price. If the contractor's bid price is increased, the owner may either accept the bid or disqualify it if it is no longer low.

When this procedure is not followed, a statement should be included providing that the contractor shall not enter into a contract with any subcontractor before approval by the owner and, in turn, that the contractor will not be required to use any subcontractor against whom he has a reasonable objection.

SUBCONTRACTUAL RELATIONS

The principal aim here is to assure that the contractor binds his subcontractors to the terms of the Contract Documents as far as they are applicable to the work of individual subcontractors. As mentioned earlier, it has long been customary to refer to the General Conditions and Supplementary Conditions at the head of each section of the Specifications, pointing out that the provisions of these documents apply with full force to the work described in them. It should be made the contractor's responsibility to obligate the subcontractor to recognize and agree to this situation as described in the General Conditions. Further discussion of contractor-subcontractor relationships is contained in Chapter 8.

Performance and Payment Bonds

The purpose of these types of surety bonds is to protect the owner against default on the part of the contractor and to assure that the project will be completed in accordance with the Contract Documents and turned over to the owner free and clear of liens or other encumbrances. Although it is common practice to require the contractor to furnish performance and payment bonds, especially in connection with competitive-bid contracts, it is the owner's decision whether to incorporate this provision in the contract in any particular instance. However,before electing to proceed without such bonds the matter should be explored thoroughly by the owner and his lawyer.

In many instances of public construction where the owner is a government agency, school district, or similar public body there may be statutory requirements controlling bonding procedures on construction work. In these situations both the amount and form of bond may be prescribed. When freedom of choice exists, both a performance bond and a labor and material payment bond, each in the amount of the full contract price, are frequently called for. If the bidding requirements stipulate such bonds, their cost is included in the bid and consequently paid by the contractor. If the bond requirement is imposed subsequent to receipt of bids, the cost is borne by the owner. The General Conditions usually contain an article reserving to the owner the right to require performance and payment bonds; when required, the form and amount of the bonds are specified in the Supplementary Conditions. A specific provision should require the contractor to submit evidence of bonding before starting any work.

Surety bonds are discussed in more detail in Chapter 6, but it may be noted here that, although a bond is not a substitute for competency and integrity on the part of the contractor, it does reinforce the financial resources of the contractor insofar as the owner is concerned.

Royalties and Patents

The expanding development of proprietary building systems makes the possibility of patent-rights infringement of increasing concern to owners, contractors, and architects. When a proprietary design or process is knowingly specified by the architect-engineer, the fact that a royalty or license fee may be involved in its use should be noted in the Specifications. This fee will then be reflected in the contractor's bid. If, however, the requirement for payment of a royalty or license fee is not cited in the Specifications, the owner and/or contractor may become liable to claims for infringement; it is then possible that the architect as the owner's agent may share any such liability. There is also the possibility that the contractor may employ some proprietary construction process or technique in his own operations that might not be specifically required or implied in the Specifications. For these reasons, it is customary to include an article on the subject in the General Conditions.

This article usually provides that the contractor shall pay all royalties and license fees, that he shall defend suits and claims for infringement, and shall hold the owner free from any costs resulting. Exceptions are instances where the claims arise in connection with systems or components specified without knowledge that they are proprietary, in which case the owner becomes responsible for the cost unless the contractor, having reason to believe an infringement might be involved, did not promptly notify the architect.

Arbitration of Claims and Disputes

The factors most prolific in generating disputes under construction contracts are claims for extra costs and extensions of time by the contractor. These arise in connection with change orders, foundation conditions at variance with those delineated in the Drawings and Specifications, delays resulting from acts of the owner or to conditions beyond the contractor's control, and conflicting requirements between Drawings and Specifications as well as errors and omissions in these documents. As noted in the earlier discussion of the architect's administrative functions, claims and disputes between contractor and owner relating to execution and progress of the work or interpretation of the Contract Documents are to be referred to the architect initially. If the architect's decision is unacceptable to either of the parties and subsequent negotiation between owner and contractor with the architect's assistance does not resolve the issue, the only alternatives are arbitration or litigation.

Because so many disputes arising under construction contracts are technical, arbitration by arbitrators familiar with the customs and practices of the building industry offers advantages of speed and economy over formal litigation. At the same time administered arbitration conducted under the Construction Industry Arbitration Rules of the American Arbitration Association promotes the reaching of fair and reasonable decisions and awards. As noted under the discussion of arbitration related to contracts for architectural and engineering services in Chapter 2, the enforceability of agreements to arbitrate varies among the states; but where they are enforceable and one party will not accept the decision of the arbitrators, judgment upon the award may be entered in a court of proper jurisdiction and enforced as any other judgment.

Care must be exercised in framing the arbitration provisions of the General Conditions so that the pertinent articles will be consonant with the laws of the state which will govern the contract. As pointed out earlier, some of these statutes prescribe rules and conditions governing arbitration procedures that must be strictly complied with if awards of the arbitrators are to be enforceable. The arbitration article should also include a clause requiring the contractor to continue the work and maintain the progress schedule during any arbitration proceedings.

INITIATION OF ARBITRATION

The General Conditions should provide that a demand for arbitration of a dispute may not be made before the date the architect has rendered his decision in the matter, or until a stipulated number of days have elapsed after the question was presented to the architect without his rendering a decision. In this situation, if the architect renders decision after arbitration proceedings have been initiated, his decision will not supersede those proceedings unless it happens to be acceptable to both parties. On the other hand, in order to cover the situation where the decision is within the allotted time but is not acceptable to one of the parties, a period should be stipulated within which a demand for arbitration must be made if it is to be made at all. Failure to demand arbitration within this period should result in the architect's decision becoming binding on both parties.

When a construction contract contains an arbitration provision to be implemented through the Construction Industry Arbitration Rules of the American Arbitration Association, one of the parties initiates action by filing with the other party a notice of his intention to arbitrate. This notice, called the Demand, contains a statement concerning the nature of the dispute, the amount of money involved, and the remedy sought. Copies of the Demand are filed with the American Arbitration Association (AAA) and with the architect. From this point all steps in the arbitration process follow the AAA's rules of procedure which are given in full in Appendix C. Examination shows that after filing of the Demand the principal subsequent events are the appointment of arbitrators, the hearings, and the award.

APPOINTMENT OF ARBITRATORS

Appointment may be accomplished in three ways: from the National Panel of Construction Arbitrators maintained by the AAA; directly by the parties concerned; or a neutral arbitrator by arbitrators selected unilaterally by each party (known as party-appointed arbitrators). Procedures for effecting appointments under each of these methods are given in the AAA rules. The appointed arbitrators constitute the arbitration Tribunal.

HEARINGS

In conducting the hearings, the arbitrators must permit full and equal opportunity to each party for the presentation of his case including witnesses, proofs, and other evidence. All witnesses must submit to questions or other examination. The parties may be represented by counsel if desired and may submit any evidence they wish. The arbitrators are the judge of evidence offered. It should be noted that conformity to the legal rules of evidence is not required in arbitration hearings.

AWARD

The arbitrators may grant any remedy or relief deemed equitable and just within the scope of the Demand and the terms of the construction contract between the parties. The award must be in writing, signed by at least a majority of the arbitrators and rendered within the agreed-upon time with a copy delivered to each of the parties.

LIMITS ON ARBITRATION

It should be noted that not all controversies arising from a construction operation are settled by arbitration. Disputes involving personal injuries or damage to property of third persons are generally settled by the courts because there is no contractual relationship between the parties. Although arbitration awards are enforceable at law once the parties have participated in the arbitration, subsequent judicial proceedings may vacate an award if it can be shown that the arbitrators were acting beyond their authority, were guilty of partiality or collusion, or that procedural requirements stipulated in the arbitration law were not followed. However, the existence of legal procedures for vacating, correcting, or confirming awards underlines the fact that parties to a dispute may avail themselves of the advantages of arbitration and still retain the safeguards of the law.

Except for acknowledging under certain conditions the validity of awards resulting from arbitration of disputes between prime contractors and subcontractors, private arbitration is not an avenue for settlement of controversies in construction contracts with the federal government.

Disputes Clauses in Government Contracts

Controversies arising under federal government construction contracts are settled under a standard disputes clause which provides that disputes concerning questions of fact "... shall be decided by the Contracting Officer. ..." This is similar to the provisions for handling disputes under government contracts for professional services (A & E) discussed in Chapter 2. The Contracting Officer is defined as the person executing the contract on behalf of the government and includes his authorized representative. The Contracting Officer's decision may be appealed to the head of the agency involved and his decision (or that of a duly authorized representative or board established for handling such appeals) is final.

Although the government as "owner" is both a party to the contract and the judge of its performance, certain safeguards exist for the construction contractor. The disputes clause hedges the "finality" of the agency head's decision by providing that it is conclusive " ... unless the same is fraudulent or capricious or arbitrary or so grossly erroneous as necessarily to imply bad faith or is not supported by substantial evidence." These exceptions provide the basis on which decisions may be subjected to judicial review as a last resort for possible relief to the contractor.[6]

Contract Time, Delays, and Extensions

The General Conditions should stipulate that all time limits stated in the Contract Documents are of the essence of the contract. In addition, there should

[6]Additional safeguards are provided when questions of law are involved in a dispute. See the article entitled "Disputes" in GSA Form 23-A, General Provisions (Construction Contract) given at the end of this chapter.

be definitions of contract time (period established elsewhere in the Construction Documents for completion of the work), date of commencement of the work, and the date of substantial completion. The last item, as defined in AIA Document A201, is " . . . the Date certified by the Architect when construction is sufficiently complete, in accordance with the Contract Documents, so the Owner may occupy the Work . . . for the use for which it is intended." This is a most important date since, when it has been reached, the architect prepares a certificate of substantial completion delineating the responsibilities of contractor and owner for maintenance, heat, utilities, security, and insurance. The date of substantial completion also serves as a reference point for the issuing of a certificate of occupancy by the city building department or other agency having jurisdiction over such matters.

DELAYS AND TIME EXTENSIONS

Since time limits are stated to be of the essence of the contract, it should be stipulated that the contract time may be extended only by official change order. Time extensions may be justified when the contractor experiences delay caused by labor disputes, fire, extremely unseasonable weather conditions, acts or neglect on the part of owner or architect or a separate contractor employed by the owner, or other causes beyond the contractor's control. A provision should also be included establishing a maximum period following the occurrence within which the contractor must file his claim.

The provisions covering delays and extensions of time are of major importance when a fixed completion date has been established and liquidated damages may be assessed under terms of the Agreement as well as where the work is being accomplished under the separate contracts system.

Payments to the Contractor

Periodic payments are normally made to the contractor as the work progresses. Under lump-sum contracts these are based on a *schedule of values* prepared immediately after execution of the contract and before the contractor makes his first application for payment. This schedule is broken down into various segments of the work such as excavation; concrete for foundations, slabs, columns, beams, etc.; reinforcing steel and structural steel; masonry; and carpentry and millwork. The amount of money allocated for each item is listed including its proper share of overhead, profit, and other general charges, with the total being equal to the contract sum. Since the schedule includes all subcontracted work as well as that to be accomplished by the contractor's own forces, the various categories should be broken down to facilitate payments to subcontractors. When approved by the architect-engineer, the schedule is used as a basis for the contractor's applications for payment.

PROGRESS PAYMENTS

The timing of these periodic payments whether monthly or otherwise is usually stated in the Agreement. The General Conditions should stipulate a minimum number of days before the due date of each progress payment when the contractor must submit his itemized application for payment. Since different owners may hold different views, it should be clearly indicated at the time the contract is executed just what substantiating data (such as vouchers and release of liens) will be required to accompany the contractor's payment application. This is particularly important where payments are to encompass material and components delivered to the site or other agreed location but not yet incorporated in the work. In this situation it is necessary to assure that title to the items thus paid for does in fact pass to the owner.

When requirements for filing the application have been satisfied, the architect issues a certificate for payment. This is usually issued in the amount of 90% of the work completed during the payment period in question, 10% being retained by the owner for his protection. Often this retainage is applied only to the first half of the contract sum, subsequent progress payments being made in full. Specific requirements concerning retainage on a particular project should appear in the Agreement, including a provision that the retained amount will constitute part of the final payment upon acceptance of the work by the owner.

CERTIFICATES FOR PAYMENT

It is of utmost importance that both owner and contractor understand just what a certificate for payment represents; for this purpose a statement of the architect's function in issuing such certificates should appear in the General Conditions. From the owner's and architect's point of view it is essential that this statement be consistent with the terms and conditions of the A & E contract under which they are operating.[7] It is also important that the contractor be aware of these terms insofar as they affect the status of certificates for payment. AIA Document A201 given at the end of this chapter contains an exceptionally clear and complete statement on certificates of payment. Two points of special significance should be noted: (a) that issuance of a certificate covering a progress payment does not constitute acceptance of any work not in accordance with the Contract Documents and (b) neither does it indicate that the architect has made exhaustive or continuous on-site inspections to check the work performed.

Although it is not the architect's responsibility when issuing a certificate to ascertain whether the contractor has, in fact, used the monies previously paid him to make payments to subcontractors, material suppliers, or labor, evidence

[7]See Chapter 2 in this connection: the article on Services Subsequent to Award of Construction Contract.

of failure to make such payments constitutes grounds for withholding a certificate. Among other reasons justifying the withholding of a certificate in whole or in part are defective work not remedied and reasonable doubt that the project can be completed for the unpaid balance of the contract sum or that it can be finished within the contract time.

If, through no fault of the contractor, the architect should fail to issue a certificate of payment or if the owner should fail to pay the contractor any amount certified by the architect within the time limits stipulated, the contractor may suspend work.

FINAL PAYMENT

The process leading to final completion and final payment begins when the contractor requests the architect-engineer to issue a certificate of substantial completion. At this time, a "punch list" of items yet to be completed or corrected is prepared and an inspection made. This leads to the issuing of the certificate of substantial completion as noted in the preceding article dealing with contract time. It should be stated specifically that failure to include any items on this list does not relieve the contractor of responsibility for completing all work in accordance with the Contract Documents. When the items on the list have been accomplished, the contractor requests final inspection and presents his final application for payment. If the work is found acceptable and complete, the architect issues the final certificate for payment which includes release of the retainage. Actual payment under the final certificate, however, is made contingent upon the furnishing of additional documents by the contractor as follows.

At time of final payment the contractor must submit releases or waivers of liens from all subcontractors, material vendors, and other parties to the construction; an affidavit that the releases and waivers furnished include all labor, materials, and equipment for which a lien could be filed; a consent to final payment by the surety underwriting the performance and payment bonds; and any other data substantiating payment of obligations that the owner may require. Items in this last category should be named in the Supplementary Conditions or in the General Requirements division of the Specifications.

Provision is normally included in the General Conditions to cover cases where final completion of the work is materially delayed through no fault of the contractor. This permits payment to be made for that portion of the work that is completed and accepted but does not terminate the contract.

The making of final payment by the owner constitutes a waiver of all claims by him except those arising from unsettled liens, terms of special guarantees and warranties required by the contract, and latent defects attributable directly to the contractor's failure to follow plans and specifications. Acceptance of final payment by the contractor constitutes a waiver of all claims by him except those previously presented but still unresolved.

Safety Precautions and Insurance

The General Conditions should be specific in placing the responsibility on the contractor for initiating and supervising all safety precautions and programs. Minimum standards, precautions, and programs which must be implemented for the safety of workers are generally prescribed in the codes promulgated by state industrial commissions under occupational safety legislation. In addition, the contractor is responsible for safety precautions necessary to protect the public and persons other than his own employees involved in the project. It is not a function of the architect to specify safety measures.

In addition to personnel safety measures prescribed by law, the contractor must also provide protection to prevent damage or loss to (a) work in place, (b) materials and equipment not yet incorporated but stored at the site or elsewhere, (c) other property at the site or adjacent to it such as trees, walks, lawns, and roadways not scheduled for removal during construction. Insurance coverage or loss by fire, wind, vandalism, earthquake damage, and similar risks is generally provided by the owner. If the contractor is to be made responsible for this type of insurance, the amount desired should be determined in advance and specified in the Contract Documents since this cost will be reflected in the bids.

Placing the responsibility on the contractor for safety precautions and injury or damage to persons or property by clauses in the General Conditions will not provide full protection to the owner. Since an injured party may name the owner as well as the contractor in a suit for damages, public liability insurance should also be required. This situation, as well as other aspects of construction insurance, is discussed more fully in Chapter 6.

Changes in the Work

Contracts for building construction invariably provide that the owner may order changes in the work without invalidating the contract. These may involve additive changes, deductive changes, or other revisions. The mechanism for making such changes is the *change order,* defined as a written order to the contractor signed by the owner and architect or by the architect alone if the owner has, in writing, appropriately authorized him to do so. If the owner has given such authority, he should inform the contractor since no change in the contract sum or contract time may be made without appropriate authorization under the terms of the construction contract.

When a change is desired by the owner, the architect prepares the necessary description, together with any supplementary drawings and/or specifications required. The contractor then prepares his proposal for effecting the change, indicating the additional cost involved (where the change represents an increase in the contract amount) and any increase in the contract time necessary to carry out the change. This proposal is usually made in the form of a lump sum, properly itemized, which is either accepted by the owner or modified by negotiation between owner and contractor with the assistance of the architect-engineer. In order to provide for situations where agreement on price cannot be reached, an article should be included in the General Conditions requiring the contractor to proceed with the work, keeping accurate records of his costs including reasonable allowance for overhead and profit. This itemized accounting then serves as a basis for subsequent resolution of the matter.[8]

MINOR CHANGES

Minor changes are defined as those that may be ordered by the architect on his own initiative and that do not involve adjustment of the contract sum or an extension of time. They must not be inconsistent with the intent of the Contract Documents and may be authorized by written *field order*. They are binding on both owner and contractor.

Uncovering and Correction of Work

The articles of the General Conditions pertaining to this subject are intended to assure that the architect-engineer will have opportunity to observe such portions of the work as he may deem necessary before being covered. Designation of specific portions of the work to be so observed should be made in the Supplementary Conditions or by special written instructions to the contractor well in advance of the time when the work would normally be covered. When work is covered prematurely contrary to such instructions, it must be uncovered if the architect so orders, and the cost of uncovering and replacement borne by the contractor even though the work is found to be in accordance with the Contract Documents. On the other hand, if the architect requires the uncovering of work not specifically designated for prior observation as noted above, the contractor is not held responsible for the cost of uncovering and replacement unless the completed work is not in accordance with the Contract Documents. If such work is found satisfactory, the costs involved will be borne by the owner through issuance of a change order.

The contractor is required to correct all defective work at his own expense whether observed before the date of substantial completion or within one year thereafter. If in accomplishing such corrections the work of other (separate) contractors is damaged, he must also bear the cost of making good that work. Time limits and procedures involved in the correction of defective work should

---

[8]It should be noted that it is not customary to rebate any portion of the original overhead and profit allowance in a deductive change order proposal.

be specified in considerable detail. It would be worthwhile to compare provisions on this subject contained in the two documents (AIA Document A201 and GSA Form 23-A) at the end of this chapter.

The owner usually reserves the right to accept defective or nonconforming work instead of requiring correction. Such acceptance constitutes a change in the contract and consequently can be ordered only by the owner. An agreed-upon reduction in the contract sum is then effected by change order.

Termination of the Contract

Provision for termination of the contract short of full performance should always be incorporated in the General Conditions. Federal government construction contracts contain a clause providing for termination at any time for convenience of the government, that is, whenever it appears in the government's interest to do so. Private contracts often have similar clauses in order to cover the owner when a project has to be abandoned because of failure to obtain financing, failure to obtain anticipated zoning changes, and for other causes. Since work completed up to the time of termination for convenience is paid for on the basis of a negotiated amount, a prearranged method of settlement should be indicated in the General Conditions. This should include procedures for (a) notifying the contractor of intent to terminate, (b) establishing valuation of work already accomplished, (c) determining contractor's reasonable profit, and (d) evaluating expenses incurred by the contractor in canceling his orders, disposing of material delivered but not yet incorporated in the project, and other costs incident to demobilizing and closing out the work.

Termination for default or breach by either contractor or owner presents a quite different situation. Implementation of the termination provisions of the contract represents an action of extreme resort and should be undertaken only with advice of legal counsel.

TERMINATION BY OWNER

It is common in private construction contracts to stipulate that the owner may terminate the contract for any of the following reasons.

1. Persistent or repeated failure or refusal by the contractor to supply enough properly skilled workmen or proper materials to carry the construction forward on schedule.

2. Failure by the contractor to make prompt payment to subcontractors or for labor or materials.

3. Persistent disregard by the contractor of applicable codes, laws, ordinances, or orders of any public authority having jurisdiction or instructions from the architect.

4. Adjudgment of the contractor as a bankrupt.

5. General assignment by the contractor for the benefit of his creditors.

6. Appointment of a receiver because of contractor's insolvency.

When the architect certifies that sufficient cause exists to justify such action, the owner gives written notice to the contractor and the surety underwriting the performance and payment bonds. He may then terminate the contractor's employment, take possession of the site, and proceed in conjunction with the surety to finish the work at the expense of the contractor or the surety. It is, of course, absolutely essential that there be no doubt about the existence of sufficient cause before instituting termination proceedings. If the cause should subsequently be held insufficient, the owner might be held liable to the contractor for damages.

TERMINATION BY CONTRACTOR

The principal grounds for suspension of work or termination of the contract on the contractor's initiative are failure to receive payment due and suspension of work for a period of thirty days (or other agreed period) by legal order not prompted by act or fault of the contractor. As noted in the discussion of payments to the contractor in a preceding article, if the architect should fail to issue a certificate of payment or if the owner should fail to pay any amount certified by the architect within the time limits stipulated, the contractor may stop work until payment has been received. When work has thus been stopped for thirty days, the contractor may, on written notice, terminate the contract and recover payment for all work executed to date plus any proven loss sustained because of the suspension, including reasonable profit and damages. When a suspension of work due to court order or the order of any other cognizant public authority exceeds thirty days, the contractor may follow the same procedure. This gives him protection analogous to that discussed in connection with termination for convenience of the owner.

Although the General Conditions are explicit with respect to the contractor's right to stop work when the owner withholds payment previously certified by the architect, considerable caution should be exercised in taking such action on the grounds that the architect has failed to issue a certificate for payment. There are several grounds on which the architect may delay issue of a certificate, and before the contractor decides to stop work he should be certain that the withholding of the certificate is not for just cause.

Miscellaneous Provisions

It should be stipulated that the contract will be governed by the law of the place where the project is being built. Since this location may be different from that of the home offices of contractor or architect-engineer, legal counsel should be consulted where different states are involved.

In view of its special significance to the owner, it is essential that the article on successors and assigns be very carefully considered. Attention is directed to the provisions covering this matter in the AIA standardized General Conditions given at the end of the chapter.

OWNER'S RIGHT TO CARRY OUT THE WORK

This provision furnishes a mechanism, short of official termination, for carrying on the work if the contractor defaults or neglects to perform any provision of the contract. After a prescribed number of days' written notice to the contractor, the owner may proceed to rectify any such deficiencies. The cost is covered by issuing a change order that deducts the amount involved from payments due or subsequently due the contractor. Such action by the owner must be approved by the architect and, in view of its possible effect on bonding or owner's liability coverage, should be undertaken only after advice of legal counsel.

TESTS AND INSPECTIONS

Requirements for quality-control and performance testing of materials and components are usually itemized in the Supplementary Conditions or the Specifications. The general rules and procedures covering tests and inspections, however, should appear in the General Conditions and extend to tests and inspections required by any public authority having jurisdiction whether or not they are identified in the Contract Documents. It is customary practice to require the contractor to bear the costs of tests in this category, but practice differs with regard to quality-control tests that require certification by an independent private testing agency. Rather than have the costs of these tests accounted for in the bid, some architect-engineers prefer that the owner pay for them directly as the job progresses, thereby eliminating possible conflict-of-interest situations between the contractor and the testing agencies. Regardless of these considerations the contractor should be assigned responsibility for giving timely notice of the readiness of the work for inspection and testing and the date arranged so that the architect-engineer and others concerned may observe the operations.

In some instances special, unspecified tests may be found necessary because of developments during the progress of construction. To cover this contingency it should be stated that upon written authorization from the owner, the architect will instruct the contractor to arrange such tests and related inspections. If such testing reveals noncompliance with either the requirements of the Contract Documents or any cognizant public authority concerning performance of the work, the contractor shall bear all costs of the special testing; otherwise the owner will bear the costs and the contractor will be reimbursed through a change order.

Separate Contracts

As noted in the early part of the chapter, this discussion of articles of the General Conditions relates directly to construction under a single general contract. When separate mechanical and electrical contracts are to be awarded in addition to a contract for the general construction work, each one should have its own General Conditions. There is no reason why substantially the same document cannot be used in each of these separate contracts; and in fact, it is highly desirable that all their provisions be as nearly alike as possible. One device that facilitates use of nearly identical General Conditions in such multiple prime contracts is the inclusion of an article reserving the right of the owner to award other contracts in connection with the project and stipulating the relationships of the separate contractors involved. It will be noted that this arrangement is incorporated in AIA Document A201 and forms the basis of the following comments.

IDENTIFICATION OF CONTRACTORS

It is necessary to clarify the meaning of the designation "contractor" in the several documents making up each separate contract. This is accomplished by stipulating that the contractor ". . . in each case shall be the contractor who signs each separate contract."

MUTUAL RESPONSIBILITY OF CONTRACTORS

Each contractor must agree to cooperate with other separate contractors with respect to storage of materials and equipment, execution of their work, and coordination of his work with theirs. Of special importance is a provision that if any portion of a contractor's work depends on the quality of another contractor's work for its execution, he shall make an inspection and report to the architect any defects that should be remedied before he proceeds further. Of equal importance is the inclusion of a clause delineating procedures to be followed in event one contractor damages the work or property of another.

CUTTING AND PATCHING

This operation can become troublesome under separate contracts if adequate cooperation is not achieved since there is usually no sharp line clearly defining responsibilities. The intention, however, should be made clear by stipulating that the contractor in question shall do all fitting, cutting, or patching of his work that is required to fit it to the work of other contractors.

CLEANING UP

In order to assure that the premises will be kept reasonably free from rubbish and waste materials accumulation during construction and that the job will be cleaned up on completion, the owner should reserve the right to clean up if disputes arise among the separate contractors as to their responsibilities in this connection. Such a clause should provide that the owner may clean up and charge the cost to the several contractors, prorated in accordance with the architect-engineer's judgment.

COORDINATION AND SCHEDULING

None of the provisions discussed touches on this key problem posed by construction under separate contracts. Since general management of the operations must be established, it must be provided either by the architect-engineer's organization or by the owner if he possesses his own in-house construction management capability. The manner in which this situation is to be handled should be clearly stipulated in the separate Agreements or in the Supplementary Conditions, since coordination of the work is of major importance with respect to time of completion and liquidated damages stipulations.

The Supplementary Conditions

When the Conditions of the Contract are written to apply solely to a specific project, there is no need for separate documents covering General Conditions and Supplementary Conditions. However, in order to realize the many advantages of employing published General Conditions, recognized as standards within the building industry, some mechanism is necessary for modifying and extending the standardized provisions. This is supplied by the Supplementary Conditions.

MODIFICATIONS

When a standard provision is at variance with the requirements of a specific project, it must be annulled, amended, or added to. Great care must be exercised in framing a modification to assure that its language is coordinated with and does not contradict the language of the standard document nor weaken its intent. In general, modifications prepared by the architect-engineer should consist only of adjustments called for by requirements of geographic location, project administrative procedures, and project complexity. If changes in other than these areas appear necessary, legal counsel should be consulted.

Articles of the General Conditions that normally need additional spelling out of requirements in the Supplementary Conditions are (among others) those relating to number of sets of Contract Documents to be furnished the contractor; identification of surveys furnished by the owner; special guarantees covering equipment or components; nature of the contractor's progress schedule (that is, bar chart or critical path network) and number of days allowed for preparing it; number of days allowed for submitting list of proposed subcontractors; percentage of progress payments to be retained; dollar amounts of the required insurance policies; and special data required by owner at time of final payment.

ADDITIONAL ARTICLES

When a particular project requires additional provisions of a contractual-legal nature, these are often incorporated as additional articles of the Supplementary Conditions. Practice varies in this regard, however, and special stipulations concerning completion of the work and those relating to liquidated damages are frequently made a part of the Agreement. Mandatory wage schedules and requirements circumscribing sources from which certain materials may be purchased usually appear in the Supplementary Conditions. At one time certain provisions pertaining to the work as a whole, such as field offices, barricades, progress photographs, and watchmen, were also included, but this custom is declining. The current trend is toward placing such contractor-furnished items and operating procedures pertaining to the job as a whole in the Specifications under the division covering nontechnical provisions and general requirements.[9] Also, nothing should appear in the Supplementary Conditons that can be covered appropriately in the technical divisions of the Specifications.

FORMAT

The principal concern here is with a system for numbering the articles of the Supplementary Conditions and cross-referencing them to the applicable articles of the General Conditions. Regardless of the numbering system, however, it is recommended that cross-referencing be done within the text of each particular article of the Supplementary Conditions. The article numbering can then be independent of the cross-referencing. Some architect-engineers start a new numbering system for the Supplementary Conditons, beginning with "Article 1," whereas others continue the sequence of the standard General Conditions. (Thus if AIA Document A201 were being used, the first article of the Supplementary Conditions would be "Article 15," since A201 contains 14 articles.) This method has the merit of eliminating any possible confusion that might arise from the two sets of articles of the Conditions of the Contract being numbered similarly.

EXAMPLES OF STANDARD FORMS

The widely used standard forms pertaining to the Conditions of the Contract reproduced here for illustrative purposes are AIA Document A201 and Form 23-A of the General Services Administration. The former is a copyrighted document and reproduction is by permission of the American Institute of Architects.

[9]This is Division 1 of the Uniform System for Construction Specifications discussed in the following chapter.

THE AMERICAN INSTITUTE OF ARCHITECTS

AIA Document A201

General Conditions of the Contract for Construction

TABLE OF ARTICLES

AIA DOCUMENT A 201 • GENERAL CONDITIONS OF THE CONTRACT FOR CONSTRUCTION • ELEVENTH EDITION • AIA ®
SEPTEMBER 1967 © THE AMERICAN INSTITUTE OF ARCHITECTS, 1735 NEW YORK AVENUE, N.W., WASHINGTON, D.C. 20006

This document is copyrighted by The American Institute of Architects and is reproduced here with its permission.

INDEX

GENERAL CONDITIONS OF THE CONTRACT FOR CONSTRUCTION

ARTICLE 1

CONTRACT DOCUMENTS

1.1   DEFINITIONS

1.1.1   THE CONTRACT DOCUMENTS

The Contract Documents consist of the Agreement, the Conditions of the Contract (General, Supplementary and other Conditions), the Drawings, the Specifications, all Addenda issued prior to execution of the Agreement, and all Modifications thereto. A Modification is (1) a written amendment to the Contract signed by both parties, (2) a Change Order, (3) a written interpretation issued by the Architect pursuant to Subparagraph 1.2.5, or (4) a written order for a minor change in the Work issued by the Architect pursuant to Paragraph 12.3. A Modification may be made only after execution of the Contract.

1.1.2   THE CONTRACT

The Contract Documents form the Contract. The Contract represents the entire and integrated agreement between the parties hereto and supersedes all prior negotiations, representations, or agreements, either written or oral, including the bidding documents. The Contract may be amended or modified only by a Modification as defined in Subparagraph 1.1.1.

1.1.3   THE WORK

The term Work includes all labor necessary to produce the construction required by the Contract Documents, and all materials and equipment incorporated or to be incorporated in such construction.

1.1.4   THE PROJECT

The Project is the total construction designed by the Architect of which the Work performed under the Contract Documents may be the whole or a part.

1.2   EXECUTION, CORRELATION, INTENT AND INTERPRETATIONS

1.2.1   The Contract Documents shall be signed in not less than triplicate by the Owner and Contractor. If either the Owner or the Contractor or both do not sign the Conditions of the Contract, Drawings, Specifications, or any of the other Contract Documents, the Architect shall identify them.

1.2.2   By executing the Contract, the Contractor represents that he has visited the site, familiarized himself with the local conditions under which the Work is to be performed, and correlated his observations with the requirements of the Contract Documents.

1.2.3   The Contract Documents are complementary, and what is required by any one shall be as binding as if required by all. The intention of the Documents is to include all labor, materials, equipment and other items as provided in Subparagraph 4.4.1 necessary for the proper execution and completion of the Work. It is not intended that Work not covered under any heading, section, branch, class or trade of the Specifications shall be supplied unless it is required elsewhere in the Contract Documents or is reasonably inferable therefrom as being necessary to produce the intended results. Words which have well-known technical or trade meanings are used herein in accordance with such recognized meanings.

1.2.4   The organization of the Specifications into divisions, sections and articles, and the arrangement of Drawings shall not control the Contractor in dividing the Work among Subcontractors or in establishing the extent of Work to be performed by any trade.

1.2.5   Written interpretations necessary for the proper execution or progress of the Work, in the form of drawings or otherwise, will be issued with reasonable promptness by the Architect and in accordance with any schedule agreed upon. Such interpretations shall be consistent with and reasonably inferable from the Contract Documents, and may be effected by Field Order.

1.3   COPIES FURNISHED AND OWNERSHIP
1.3.1   Unless otherwise provided in the Contract Documents, the Contractor will be furnished, free of charge, all copies of Drawings and Specifications reasonably necessary for the execution of the Work.

1.3.2   All Drawings, Specifications and copies thereof furnished by the Architect are and shall remain his property. They are not to be used on any other project, and, with the exception of one contract set for each party to the Contract, are to be returned to the Architect on request at the completion of the Work.

ARTICLE 2

ARCHITECT

2.1   DEFINITION
2.1.1   The Architect is the person or organization identified as such in the Agreement and is referred to throughout the Contract Documents as if singular in number and masculine in gender. The term Architect means the Architect or his authorized representative.

2.1.2   Nothing contained in the Contract Documents shall create any contractual relationship between the Architect and the Contractor.

2.2   ADMINISTRATION OF THE CONTRACT
2.2.1   The Architect will provide general Administration of the Construction Contract, including performance of the functions hereinafter described.

2.2.2   The Architect will be the Owner's representative during construction and until final payment. The Architect will have authority to act on behalf of the Owner to the extent provided in the Contract Documents, unless otherwise modified by written instrument which will be shown to the Contractor. The Architect will advise and consult with the Owner, and all of the Owner's instructions to the Contractor shall be issued through the Architect.

2.2.3   The Architect shall at all times have access to the Work wherever it is in preparation and progress. The Contractor shall provide facilities for such access so the Architect may perform his functions under the Contract Documents.

2.2.4   The Architect will make periodic visits to the site to familiarize himself generally with the progress and quality of the Work and to determine in general if the Work is proceeding in accordance with the Contract Documents. On the basis of his on-site observations as an architect, he will keep the Owner informed of the progress of the Work, and will endeavor to guard the Owner against defects and deficiencies in the Work of the Contractor. The Architect will not be required to make exhaustive or continuous on-site inspections to check the quality or quantity of the Work. The Architect will not be responsible for construction means, methods, techniques, sequences or procedures, or for safety precautions and programs in connection with the Work, and he will not be responsibile for the Contractor's failure to carry out the Work in accordance with the Contract Documents.

2.2.5   Based on such observations and the Contractor's Applications for Payment, the Architect will determine the amounts owing to the Contractor and will issue Certificates for Payment in such amounts, as provided in Paragraph 9.4.

2.2.6   The Architect will be, in the first instance, the interpreter of the requirements of the Contract Documents and the judge of the performance thereunder by both the Owner and Contractor. The Architect will within a reasonable time, render such interpretations as he may deem necessary for the proper execution or progress of the Work.

2.2.7   Claims, disputes and other matters in question between the Contractor and the Owner relating to the execution or progress of the Work or the interpretation of the Contract Documents shall be referred initially to the Architect for decision which he will render in writing within a reasonable time.

2.2.8   All interpretations and decisions of the Architect shall be consistent with the intent of the Contract Documents. In his capacity as interpreter and judge, he will exercise his best efforts to insure faithful performance by both the Owner and the Contractor and will not show partiality to either.

2.2.9   The Architect's decisions in matters relating to artistic effect will be final if consistent with the intent of the Contract Documents.

2.2.10   Any claim, dispute or other matter that has been referred to the Architect, except those relating to artistic effect as provided in Subparagraph 2.2.9 and except any which have been waived by the making or acceptance of final payment as provided in Subparagraphs 9.7.5 and 9.7.6, shall be subject to arbitration upon the written demand of either party. However, no demand for arbitration of any such claim, dispute or other matter may be made until the earlier of:

.1   the date on which the Architect has rendered his decision, or
.2   the tenth day after the parties have presented their evidence to the Architect or have been given a reasonable opportunity to do so, if the Architect has not rendered his written decision by that date.

2.2.11   If a decision of the Architect is made in writing and states that it is final but subject to appeal, no demand for arbitration of a claim, dispute or other matter covered by such decision may be made later than thirty days after the date on which the party making the demand received the decision. The failure to demand arbitration within said thirty days' period will result in the Architect's decision becoming final and binding upon the Owner and the Contractor. If the Architect renders a decision after arbitration proceedings have been initiated, such decision may be entered as evidence but will not supersede any arbitration proceedings except where the decision is acceptable to the parties concerned.

2.2.12   The Architect will have authority to reject Work which does not conform to the Contract Documents. Whenever, in his reasonable opinion, he considers it necessary or advisable to insure the proper implementation of the intent of the Contract Documents, he will have authority to require the Contractor to stop the Work or any portion thereof, or to require special inspection or testing of the Work as provided in Subparagraph 7.8.2 whether or not such Work be then fabricated, installed or completed. However, neither the Architect's authority to act under this Subparagraph 2.2.12, nor any decision made by him in good faith either to exercise or not to exercise such authority, shall give rise to any duty or responsibility of the Architect to the Contractor, any Subcontractor, any of their agents or employees, or any other person performing any of the Work.

2.2.13   The Architect will review Shop Drawings and Samples as provided in Subparagraphs 4.13.1 through 4.13.8 inclusive.

2.2.14   The Architect will prepare Change Orders in accordance with Article 12, and will have authority to order minor Changes in the Work as provided in Subparagraph 12.3.1.

2.2.15   The Architect will conduct inspections to determine the dates of Substantial Completion and final completion, will receive written guarantees and related documents required by the Contract and assembled by the Contractor, and will issue a final Certificate for Payment.

2.2.16   If the Owner and Architect agree, the Architect will provide one or more full-time Project Representatives to assist the Architect in carrying out his responsibilities at the site. The duties, responsibilities and limitations of authority of any such Project Representative shall be as set forth in an exhibit to be incorporated in the Contract Documents.

2.2.17   The duties, responsibilities and limitations of authority of the Architect as the Owner's representative during construction as set forth in Articles 1 through 14 inclusive of these General Conditions will not be modified or extended without written consent of the Owner and the Architect which will be shown to the Contractor.

2.2.18   The Architect will not be responsible for the acts or omissions of the Contractor, any Subcontractors, or any of their agents or employees, or any other persons performing any of the Work.

2.2.19   In case of the termination of the employment of the Architect, the Owner shall appoint an architect against whom the Contractor makes no reasonable objection, whose status under the Contract Documents shall be that of the former architect. Any dispute in connection with such appointment shall be subject to arbitration.

ARTICLE 3

OWNER

3.1   DEFINITION
3.1.1   The Owner is the person or organization identified as such in the Agreement and is referred to throughout the Contract Documents as if singular in number and masculine in gender. The term Owner means the Owner or his authorized representative.

3.2   INFORMATION AND SERVICES REQUIRED OF THE OWNER
3.2.1   The Owner shall furnish all surveys describing the physical characteristics, legal limits and utility locations for the site of the Project.

3.2.2   The Owner shall secure and pay for easements for permanent structures or permanent changes in existing facilities.

3.2.3   Information or services under the Owner's control shall be furnished by the Owner with reasonable promptness to avoid delay in the orderly progress of the Work.

3.2.4   The Owner shall issue all instructions to the Contractor through the Architect.

3.2.5   The foregoing are in addition to other duties and responsibilities of the Owner enumerated herein and especially those in respect to Payment and Insurance in Articles 9 and 11 respectively.

ARTICLE 4

CONTRACTOR

4.1   DEFINITION

4.1.1   The Contractor is the person or organization identified as such in the Agreement and is referred to throughout the Contract Documents as if singular in number and masculine in gender. The term Contractor means the Contractor or his authorized representative.

4.2   REVIEW OF CONTRACT DOCUMENTS

4.2.1   The Contractor shall carefully study and compare the Agreement, Conditions of the Contract, Drawings, Specifications, Addenda and Modifications and shall at once report to the Architect any error, inconsistency or omission he may discover; but the Contractor shall not be liable to the Owner or the Architect for any damage resulting from any such errors, inconsistencies or omissions. The Contractor shall do no Work without Drawings, Specifications or interpretations.

4.3   SUPERVISION AND CONSTRUCTION PROCEDURES

4.3.1   The Contractor shall supervise and direct the Work, using his best skill and attention. He shall be solely responsible for all construction means, methods, techniques, sequences and procedures and for coordinating all portions of the Work under the Contract.

4.4   LABOR AND MATERIALS

4.4.1   Unless otherwise specifically noted, the Contractor shall provide and pay for all labor, materials, equipment, tools, construction equipment and machinery, water, heat, utilities, transportation, and other facilities and services necessary for the proper execution and completion of the Work.

4.4.2   The Contractor shall at all times enforce strict discipline and good order among his employees and shall not employ on the Work any unfit person or anyone not skilled in the task assigned to him.

4.5   WARRANTY

4.5.1   The Contractor warrants to the Owner and the Architect that all materials and equipment furnished under this Contract will be new unless otherwise specified, and that all Work will be of good quality, free from faults and defects and in conformance with the Contract Documents. All Work not so conforming to these standards may be considered defective. If required by the Architect, the Contractor shall furnish satisfactory evidence as to the kind and quality of materials and equipment

4.5.2   The warranty provided in this Paragraph 4.5 shall be in addition to and not in limitation of any other warranty or remedy required by law or by the Contract Documents.

4.6   TAXES
4.6.1   The Contractor shall pay all sales, consumer, use and other similar taxes required by law.

4.7   PERMITS, FEES AND NOTICES
4.7.1   The Contractor shall secure and pay for all permits, governmental fees and licenses necessary for the proper execution and completion of the Work.

4.7.2   The Contractor shall give all notices and comply with all laws, ordinances, rules, regulations and orders of any public authority bearing on the performance of the Work. If the Contractor observes that any of the Contract Documents are at variance therewith in any respect, he shall promptly notify the Architect in writing, and any necessary changes shall be adjusted by appropriate Modificaiton. If the Contractor performs any Work knowing it to be contrary to such laws, ordinances, rules and regulations, and without such notice to the Architect, he shall assume full responsibility therefor and shall bear all costs attributable thereto.

4.8   CASH ALLOWANCES
4.8.1   The Contractor shall include in the Contract Sum all allowances stated in the Contract Documents. These allowances shall cover the net cost of the materials and equipment delivered and unloaded at the site, and all applicable taxes. The Contractor's handling costs on the site, labor, installation costs, overhead, profit and other expenses contemplated for the original allowance shall be included in the Contract Sum and not in the allowance. The Contractor shall cause the Work covered by these allowances to be performed for such amounts and by such persons at the Architect may direct, but he will not be required to employ persons against whom he makes a reasonable objection. If the cost, when determined, is more than or less than the allowance, the Contract Sum shall be adjusted accordingly by Change Order which will include additional handling costs on the site, labor, installation costs, overhead, profit and other expenses resulting to the Contractor from any increase over the original allowance.

4.9   SUPERINTENDENT
4.9.1   The Contractor shall employ a competent superintendent and necessary assistants who shall be in attendance at the Project site during the progress of the Work. The superintendent shall be satisfactory to the Architect, and shall not be changed except with the consent of the Architect, unless the superintendent proves to be unsatisfactory to the Contractor and ceases to be in his employ. The superintendent shall represent the Contractor and all communications given to the superintendent shall be as binding as if given to the Contractor. Important communications will be confirmed in writing. Other communications will be so confirmed on written request in each case.

4.10   RESPONSIBILITY FOR THOSE PERFORMING THE WORK
4.10.1   The Contractor shall be responsible to the Owner for the acts and omissions of all his employees and all Subcontractors, their agents and employees, and all other persons performing any of the Work under a contract with the Contractor.

4.11   PROGRESS SCHEDULE
4.11.1   The Contractor, immediately after being awarded the Contract, shall prepare and submit for the Architect's approval an estimated progress schedule for the Work. The progress schedule shall be related to the entire Project to the extent required by the Contract Documents. This schedule shall indicate the dates for the starting and completion of the various stages of construction and shall be revised as required by the conditions of the Work, subject to the Architect's approval.

4.12   DRAWINGS AND SPECIFICATIONS AT THE SITE

4.12.1   The Contractor shall maintain at the site for the Owner one copy of all Drawings, Specifications, Addenda, approved Shop Drawings, Change Orders and other Modifications, in good order and marked to record all changes made during construction. These shall be available to the Architect. The Drawings, marked to record all changes made during construction, shall be delivered to him for the Owner upon completion of the Work.

4.13   SHOP DRAWINGS AND SAMPLES

4.13.1   Shop Drawings are drawings, diagrams, illustrations, schedules, performance charts, brochures and other data which are prepared by the Contractor or any Subcontractor, manufacturer, supplier or distributor, and which illustrate some portion of the Work.

4.13.2   Samples are physical examples furnished by the Contractor to illustrate materials, equipment or workmanship, and to establish standards by which the Work will be judged.

4.13.3   The Contractor shall review, stamp with his approval and submit, with reasonable promptness and in orderly sequence so as to cause no delay in the Work or in the work of any other contractor, all Shop Drawings and Samples required by the Contract Documents or subsequently by the Architect as covered by Modifications. Shop Drawings and Samples shall be properly identified as specified, or as the Architect may require. At the time of submission the Contractor shall inform the Architect in writing of any deviation in the Shop Drawings or Samples from the requirements of the Contract Documents.

4.13.4   By approving and submitting Shop Drawings and Samples, the Contractor thereby represents that he has determined and verified all field measurements, field construction criteria, materials, catalog numbers and similar data, or will do so, and that he has checked and coordinated each Shop Drawing and Sample with the requirements of the Work and of the Contract Documents.

4.13.5   The Architect will review and approve Shop Drawings and Samples with reasonable promptness so as to cause no delay, but only for conformance with the design concept of the Project and with the information given in the Contract Documents. The Architect's approval of a separate item shall not indicate approval of an assembly in which the item functions.

4.13.6   The Contractor shall make any corrections required by the Architect and shall resubmit the required number of corrected copies of Shop Drawings or new Samples until approved. The Contractor shall direct specific attention in writing or on resubmitted Shop Drawings to revisions other than the corrections requested by the Architect on previous submissions.

4.13.7   The Architect's approval of Shop Drawings or Samples shall not relieve the Contractor of responsibility for any deviation from the requirements of the Contract Documents unless the Contractor has informed the Architect in writing of such deviation at the time of submission and the Architect has given written approval to the specific deviation, nor shall the Architect's approval relieve the Contractor from responsibility for errors or omissions in the Shop Drawings or Samples.

4.13.8   No portion of the Work requiring a Shop Drawing or Sample submission shall be commenced until the submission has been approved by the Architect. All such portions of the Work shall be in accordance with approved Shop Drawings and Samples.

4.14   USE OF SITE

4.14.1   The Contractor shall confine operations at the site to areas permitted by law, ordinances, permits and the Contract Documents and shall not unreasonably encumber the site with any materials or equipment.

4.15   CUTTING AND PATCHING OF WORK
4.15.1   The Contractor shall do all cutting, fitting or patching of his Work that may be required to make its several parts fit together properly, and shall not endanger any Work by cutting, excavating or otherwise altering the Work or any part of it.

4.16   CLEANING UP
4.16.1   The Contractor at all times shall keep the premises free from accumulation of waste materials or rubbish caused by his operations. At the completion of the Work he shall remove all his waste materials and rubbish from and about the Project as well as all his tools, construction equipment, machinery and surplus materials, and shall clean all glass surfaces and leave the Work "broom-clean" or its equivalent, except as otherwise specified.

4.16.2   If the Contractor fails to clean up, the Owner may do so and the cost thereof shall be charged to the Contractor as provided in Paragraph 7.6.

4.17   COMMUNICATIONS
4.17.1   The Contractor shall forward all communiations to the Owner through the Architect.

4.18   INDEMNIFICATION
4.18.1   The Contractor shall indemnify and hold harmless the Owner and the Architect and their agents and employees from and against all claims, damages, losses and expenses including attorneys' fees arising out of or resulting from the performance of the Work, provided that any such claim, damage, loss or expense (a) is attributable to bodily injury, sickness, disease or death, or to injury to or destruction of tangible property (other than the Work itself) including the loss of use resulting therefrom, and (b) is caused in whole or in part by any negligent act or omission of the Contractor, any Subcontractor, anyone directly or indirectly employed by any of them or anyone for whose acts any of them may be liable, regardless of whether or not it is caused in part by a party indemnified hereunder.

4.18.2   In any and all claims against the Owner or the Architect or any of their agents or employees by any employee of the Contractor, any Subcontractor, anyone directly or indirectly employed by any of them or anyone for whose acts any of them may be liable, the indemnification obligation under this Paragraph 4.18 shall not be limited in any way by any limitation on the amount or type of damages, compensation or benefits payable by or for the Contractor or any Subcontractor under workmen's compensation acts, disability benefit acts or other employee benefit acts.

4.18.3   The obligations of the Contractor under this Paragraph 4.18 shall not extend to the liability of the Architect, his agents or employees arising out of (1) the preparation or approval of maps, drawings, opinions, reports, surveys, Change Orders, designs or speci-fications, or (2) the giving of or the failure to give directions or instructions by the Architect, his agents or employees provided such giving or failure to give is the primary cause of the injury or damage.

ARTICLE 5

SUBCONTRACTORS

5.1 DEFINITION

5.1.1 A Subcontractor is a person or organization who has a direct contract with the Contractor to perform any of the Work at the site. The term Subcontractor is referred to throughout the Contract Documents as if singular in number and masculine in gender and means a Subcontractor or his authorized representative.

5.1.2 A Sub-subcontractor is a person or organization who has a direct or indirect contract with a Subcontractor to perform any of the Work at the site. The term Sub-subcontractor is referred to throughout the Contract Documents as if singular in number and masculine in gender and means a Sub-subcontractor or an authorized representative thereof.

5.1.3 Nothing contained in the Contract Documents shall create any contractual relation between the Owner or the Architect and any Subcontractor or Sub-subcontractor.

5.2 AWARD OF SUBCONTRACTS AND OTHER CONTRACTS FOR PORTIONS OF THE WORK

5.2.1 As soon as practicable after bids are received and prior to the award of the Contract, the successful bidder shall furnish to the Architect in writing for acceptance by the Owner and the Architect a list of the names of the subcontractors or other persons or organizations (including those who are to furnish materials or equipment fabricated to a special design) proposed for such portions of the Work as may be designated in the bidding requirements, or, if none is so designated the names of the Subcontractors proposed for the principal portions of the Work. Prior to the award of the Contract, the Architect shall notify the successful bidder in writing if either the Owner or the Architect, after due investigation, has reasonable objection to any person or organization on such list. Failure of the Owner or Architect to make an objection to any person or organization on the list prior to the award shall constitute acceptance of such person or organization.

5.2.2 If, prior to the award of the Contract, the Owner or Architect has a reasonable and substantial objection to any person or organization on such list, and refuses in writing to accept such person or organization, the successful bidder may, prior to the award, withdraw his bid without forfeiture of bid security. If the successful bidder submits an acceptable substitute with an increase in his bid price to cover the difference in cost occasioned by such substitution, the Owner may, at his discretion, accept the increased bid price or he may disqualify the bid. If, after the award, the Owner or Architect refuses to accept any person or organization on such list, the Contractor shall submit an acceptable substitute and the Contract Sum shall be increased or decreased by the difference in cost occasioned by such substitution and an appropriate Change Order shall be issued; however, no increase in the Contract Sum shall be allowed for any such substitution unless the Contractor has acted promptly and responsively in submitting a name with respect thereto prior to the award.

5.2.3 The Contractor shall not contract with any Subcontractor or any person or organization proposed for portions of the Work designated in the bidding requirements or, if none is so designated, with any Subcontractor proposed for the principal portions of the Work who has not been accepted by the Owner and the Architect. The Contractor will not be required to contract with any subcontractor or person or organization against whom he has a reasonable objection.

5.2.4   If the Owner or the Architect requires a change of any proposed Subcontractor or person or organization previously accepted by them, the Contract Sum shall be increased or decreased by the difference in cost occasioned by such change and an appropriate Change Order shall be issued.

5.2.5   The Contractor shall not make any substitution for any Subcontractor or person or organization who has been accepted by the Owner and the Architect, unless the substitution is acceptable to the Owner and the Architect.

5.3   SUBCONTRACTUAL RELATIONS

5.3.1   All work performed for the Contractor by a Subcontractor shall be pursuant to an appropriate agreement between the Contractor and the Subcontractor (and where appropriate between Subcontractors and Sub-subcontractors) which shall contain provisions that:

.1   preserve and protect the rights of the Owner and the Architect under the Contract with respect to the Work to be performed under the subcontract so that the subcontracting thereof will not prejudice such rights;

.2   require that such Work be performed in accordance with the requirements of the Contract Documents;

.3   require submission to the Contractor of applications for payment under each subcontract to which the Contractor is a party, in reasonable time to enable the Contractor to apply for payment in accordance with Article 9;

.4   require that all claims for additional costs, extensions of time, damages for delays or otherwise with respect to subcontracted portions of the Work shall be submitted to the Contractor (via any Subcontractor or Sub-subcontractor where appropriate) in the manner provided in the Contract Documents for like claims by the Contractor upon the Owner;

.5   waive all rights the contracting parties may have against one another for damages caused by fire or other perils covered by the property insurance described in Paragraph 11.3 except such rights as they may have to the proceeds of such insurance held by the Owner as trustee under Paragraph 11.3; and

.6   obligate each Subcontractor specifically to consent to the provisions of this Paragraph 5.3.

5.4   PAYMENTS TO SUBCONTRACTORS

5.4.1   The Contractor shall pay each Subcontractor, upon receipt of payment from the Owner, an amount equal to the percentage of completion allowed to the Contractor on account of such Subcontractor's Work. The Contractor shall also require each Subcontractor to make similar payments to his subcontractors.

5.4.2   If the Architect fails to issue a Certificate for Payment for any cause which is the fault of the Contractor and not the fault of a particular Subcontractor, the Contractor shall pay that Subcontractor on demand, made at any time after the Certificate for Payment should otherwise have been issued, for his Work to the extent completed, less the retained percentage.

5.4.3   The Contractor shall pay each Subcontractor a just share of any insurance moneys received by the Contractor under Article 11, and he shall require each Subcontractor to make similar payments to his subcontractors.

5.4.4   The Architect may, on request and at his discretion, furnish to any Subcontractor, if practicable, information regarding percentages of completion certified to the Contractor on account of Work done by such Subcontractors.

5.4.5   Neither the Owner nor the Architect shall have any obligation to pay or to see to the payment of any moneys to any Subcontractor except as may otherwise be required by law.

ARTICLE 6

SEPARATE CONTRACTS

6.1   OWNER'S RIGHT TO AWARD SEPARATE CONTRACTS

6.1.1   The Owner reserves the right to award other contracts in connection with other portions of the Project under these or similar Conditions of the Contract.

6.1.2   When separate contracts are awarded for different portions of the Project, "the Contractor" in the contract documents in each case shall be the contractor who signs each separate contract.

6.2   MUTUAL RESPONSIBILITY OF CONTRACTORS

6.2.1   The Contractor shall afford other contractors reasonable opportunity for the introduction and storage of their materials and equipment and the execution of their work, and shall properly connect and coordinate his Work with theirs.

6.2.2   If any part of the Contractor's Work depends for proper execution or results upon the work of any other separate contractor, the Contractor shall inspect and promptly report to the Architect any apparent discrepancies or defects in such work that render it unsuitable for such proper execution and results. Failure of the Contractor so to inspect and report shall constitute an acceptance of the other contractor's work as fit and proper to receive his Work, except as to defects which may develop in the other separate contractor's work after the execution of the Contractor's Work.

6.2.3   Should the Contractor cause damage to the work or property of any separate contractor on the Project, the Contractor shall, upon due notice, settle with such other contractor by agreement or arbitration, if he will so settle. If such separate contractor sues the Owner on account of any damage alleged to have been so sustained, the Owner shall notify the Contractor who shall defend such proceedings at the Owner's expense, and if any judgment against the Owner arises therefrom the Contractor shall pay or satisfy it and shall reimburse the Owner for all attorneys' fees and court costs which the Owner has incurred.

6.3   CUTTING AND PATCHING UNDER SEPARATE CONTRACTS

6.3.1   The Contractor shall do all cutting, fitting or patching of his Work that may be required to fit it to receive or be received by the work of other contractors shown in the Contract Documents. The Contractor shall not endanger any work of any other contractors by cutting, excavating or otherwise altering any work and shall not cut or alter the work of any other contractor except with the written consent of the Architect.

6.3.2   Any costs caused by defective or ill-timed work shall be borne by the party responsible therefor.

6.4   OWNER'S RIGHT TO CLEAN UP

6.4.1   If a dispute arises between the separate contractors as to their responsibility for cleaning up as required by Paragraph 4.16, the Owner may clean up and charge the cost thereof to the several contractors as the Architect shall determine to be just.

ARTICLE 7

MISCELLANEOUS PROVISIONS

7.1 LAW OF THE PLACE
7.1.1   The Contract shall be governed by the law of the place where the Project is located.

7.2 SUCCESSORS AND ASSIGNS
7.2.1   The Owner and the Contractor each binds himself, his partners, successors, assigns and legal representatives to the other party hereto and to the partners, successors, assigns and legal representatives of such other party in respect to all covenants, agreements and obligations contained in the Contract Documents. Neither party to the Contract shall assign the Contract or sublet it as a whole without the written consent of the other, nor shall the Contractor assign any moneys due or to become due to him hereunder, without the previous written consent of the Owner.

7.3 WRITTEN NOTICE
7.3.1   Written notice shall be deemed to have been duly served if delivered in person to the individual or member of the firm or to an officer of the corporation for whom it was intended, or if delivered at or sent by registered or certified mail to the last business address known to him who gives the notice.

7.4 CLAIMS FOR DAMAGES
7.4.1   Should either party to the Contract suffer injury or damage to person or property because of any act or omission of the other party or of any of his employees, agents or others for whose acts he is legally liable, claim shall be made in writing to such other party within a reasonable time after the first observance of such injury or damage.

7.5 PERFORMANCE BOND AND LABOR AND MATERIAL PAYMENT BOND
7.5.1   The Owner shall have the right, prior to signing the Contract, to require the Contractor to furnish bonds covering the faithful performance of the Contract and the payment of all obligations arising thereunder in such form and amount as the Owner may prescribe and with such sureties as may be agreeable to the parties. If such bonds are stipulated in the bidding requirements, the premiums shall be paid by the Contractor; if required subsequent to the submission of quotations or bids, the cost shall be reimbursed by the Owner. The Contractor shall deliver the required bonds to the Owner not later than the date of execution of the Contract, or if the Work is commenced prior thereto in response to a notice to proceed, the Contractor shall, prior to commencement of the Work, submit evidence satisfactory to the Owner that such bonds will be issued.

7.6 OWNER'S RIGHT TO CARRY OUT THE WORK
7.6.1   If the Contractor defaults or neglects to carry out the Work in accordance with the Contract Documents or fails to perform any provision of the Contract, the Owner may, after seven days' written notice to the Contractor and without prejudice to any other remedy he may have, make good such deficiencies. In such case an appropriate Change Order shall be issued deducting from the payments then or thereafter due the Contractor the cost of correcting such deficiencies, including the cost of the Architect's additional services made necessary by such default, neglect or failure. The Architect must approve both such action and the amount charged to the Contractor. If the payments then or thereafter due the Contractor are not sufficient to cover such amount, the Contractor shall pay the difference to the Owner.

7.7 ROYALTIES AND PATENTS

7.7.1   The Contractor shall pay all royalties and license fees. He shall defend all suits or claims for infringement of any patent rights and shall save the Owner harmless from loss on account thereof, except that the Owner shall be responsible for all such loss when a particular design, process or the product of a particular manufacturer or manufacturers is specified, but if the Contractor has reason to believe that the design, process or product specified is an infringement of a patent, he shall be responsible for such loss unless he promptly gives such information to the Architect.

7.8 TESTS

7.8.1   If the Contract Documents, laws, ordinances, rules, regulations or orders of any public authority having jurisdiction require any Work to be inspected, tested or approved, the Contractor shall give the Architect timely notice of its readiness and of the date arranged so the Architect may observe such inspection, testing or approval. The Contractor shall bear all costs of such inspections, tests and approvals unless otherwise provided.

7.8.2   If after the commencement of the Work the Architect determines that any Work requires special inspection, testing or approval which Subparagraph 7.8.1 does not include, he will, upon written authorization from the Owner, instruct the Contractor to order such special inspection, testing or approval, and the Contractor shall give notice as in Subparagraph 7.8.1. If such special inspection or testing reveals a failure of the Work to comply (1) with the requirements of the Contract Documents or (2), with respect to the performance of the Work, with laws, ordinances, rules, regulations or orders of any public authority having jurisdiction, the Contractor shall bear all costs thereof, including the Architect's additional services made necessary by such failure; otherwise the Owner shall bear such costs, and an appropriate Change Order shall be issued.

7.8.3   Required certificates of inspection, testing or approval shall be secured by the Contractor and promptly delivered by him to the Architect.

7.8.4   If the Architect wishes to observe the inspections, tests or approvals required by this Paragraph 7.8, he will do so promptly and, where practicable, at the source of supply.

7.8.5   Neither the observations of the Architect in his administration of the Contract, nor inspections, tests or approvals by persons other than the Contractor shall relieve the Contractor from his obligations to perform the Work in accordance with the Contract Documents.

7.9 INTEREST

7.9.1   Any moneys not paid when due to either party under this Contract shall bear interest at the legal rate in force at the place of the Project.

7.10 ARBITRATION

7.10.1   All claims, disputes and other matters in question arising out of, or relating to, this Contract or the breach thereof, except as set forth in Subparagraph 2.2.9 with respect to the Architect's decisions on matters relating to artistic effect, and except for claims which have been waived by the making or acceptance of final payment as provided by Subparagraphs 9.7.5 and 9.7.6, shall be decided by arbitration in accordance with the Construction Industry Arbitration Rules of the American Arbitration Association then obtaining unless the parties mutually agree otherwise. This agreement so to arbitrate shall be specifically enforceable under the prevailing arbitration law. The award rendered by the arbitrators shall be final, and judgment may be entered upon it in accordance with applicable law in any court having jurisdiction thereof.

7.10.2   Notice of the demand for arbitration shall be filed in writing with the other party to the Contract and with the American Arbitration Association, and a copy shall be filed with the Architect. The demand for arbitration shall be made within the time limits specified in Subparagraphs 2.2.10 and 2.2.11 where applicable, and in all other cases within a reasonable time after the claim, dispute or other matter in question has arisen, and in no event shall it be made after institution of legal or equitable proceedings based on such claim, dispute or other matter in question would be barred by the applicable statute of limitations.

7.10.3   The Contractor shall carry on the Work and maintain the progress schedule during any arbitration proceedings, unless otherwise agreed by him and the Owner in writing.

ARTICLE 8

TIME

8.1   DEFINITIONS
8.1.1   The Contract Time is the period of time allotted in the Contract Documents for completion of the Work.

8.1.2   The date of commencement of the Work is the date established in a notice to proceed. If there is no notice to proceed, it shall be the date of the Agreement or such other date as may be established therein.

8.1.3   The Date of Substantial Completion of the Work or designated portion thereof is the Date certified by the Architect when construction is sufficiently complete, in accordance with the Contract Documents, so the Owner may occupy the Work or designated portion thereof for the use for which it is intended.

8.2   PROGRESS AND COMPLETION
8.2.1   All time limits stated in the Contract Documents are of the essence of the Contract.

8.2.2   The Contractor shall begin the Work on the date of commencement as defined in Subparagraph 8.1.2. He shall carry the Work forward expeditiously with adequate forces and shall complete it within the Contract Time.

8.3   DELAYS AND EXTENSIONS OF TIME
8.3.1   If the Contractor is delayed at any time in the progress of the Work by any act or neglect of the Owner or the Architect, or by any employee of either, or by any separate contractor employed by the Owner, or by changes ordered in the Work, or by labor disputes, fire, unusual delay in transportation, unavoidable casualties or any causes beyond the Contractor's control, or by delay authorized by the Owner pending arbitration, or by any cause which the Architect determines may justify the delay, then the Contract Time shall be extended by Change Order for such reasonable time as the Architect may determine.

8.3.2   All claims for extension of time shall be made in writing to the Architect no more than fifteen days after the occurrence of the delay; otherwise they shall be waived. In the case of a continuing cause of delay only one claim is necessary.

8.3.3   If no schedule or agreement is made stating the dates upon which written interpretations as set forth in Subparagraph 1.2.5 shall be furnished, then no claim for delay shall be allowed on account of failure to furnish such interpretations until fifteen days after demand is made for them, and not then unless such claim is reasonable.

8.3.4   This Paragraph 8.3 does not exclude the recovery of damages for delay by either party under other provisions of the Contract Documents.

ARTICLE 9

PAYMENTS AND COMPLETION

9.1   CONTRACT SUM
9.1.1   The Contract Sum is stated in the Agreement and is the total amount payable by the Owner to the Contractor for the performance of the Work under the Contract Documents.

9.2   SCHEDULE OF VALUES
9.2.1   Before the first Application for Payment, the Contractor shall submit to the Architect a schedule of values of the various portions of the Work, including quantities if required by the Architect, aggregating the total Contract Sum, divided so as to facilitate payments to Subcontractors in accordance with Paragraph 5.4, prepared in such form as specified or as the Architect and the Contractor may agree upon, and supported by such data to substantiate its correctness as the Architect may require. Each item in the schedule of values shall include its proper share of overhead and profit. This schedule, when approved by the Architect, shall be used only as a basis for the Contractor's Applications for Payment.

9.3   PROGRESS PAYMENTS
9.3.1   At least ten days before each progress payment falls due, the Contractor shall submit to the Architect an itemized Application for Payment, supported by such data substantiating the Contractor's right to payment as the Owner or the Architect may require.

9.3.2   If payments are to be made on account of materials or equipment not incorporated in the Work but delivered and suitably stored at the site, or at some other location agreed upon in writing, such payments shall be conditioned upon submission by the Contractor of bills of sale or such other procedures satisfactory to the Owner to establish the Owner's title to such materials or equipment or otherwise protect the Owner's interest including applicable insurance and transportation to the site.

9.3.3   The Contractor warrants and guarantees that title to all Work, materials and equipment covered by an Application for Payment, whether incorporated in the Project or not, will pass to the Owner upon the receipt of such payment by the Contractor, free and clear of all liens, claims, security interests or encumbrances, hereinafter referred to in this Article 9 as "liens"; and that no Work, materials or equipment covered by an Application for Payment will have been acquired by the Contractor; or by any other person performing the Work at the site or furnishing materials and equipment for the Project, subject to an agreement under which an interest therein or an encumbrance thereon is retained by the seller or otherwise imposed by the Contractor or such other person.

9.4   CERTIFICATES FOR PAYMENT
9.4.1   If the Contractor has made Application for Payment as above, the Architect will, with reasonable promptness but not more than seven days after the receipt of the Application, issue a Certificate for Payment to the Owner, with a copy to the Contractor, for such amount as he determines to be properly due, or state in writing his reasons for withholding a Certificate as provided in Subparagraph 9.5.1.

9.4.2 The issuance of a Certificate for Payment will constitute a representation by the Architect to the Owner, based on his observations at the site as provided in Subparagraph 2.2.4 and the data comprising the Application for Payment, that the Work has progressed to the point indicated; that, to the best of his knowledge, information and belief, the quality of the Work is in accordance with the Contract Documents (subject to an evaluation of the Work as a functioning whole upon Substantial Completion, to the results of any subsequent tests required by the Contract Documents, to minor deviations from the Contract Documents correctable prior to completion, and to any specific qualifications stated in his Certificate); and that the Contractor is entitled to payment in the amount certified. In addition, the Architect's final Certificate for Payment will constitute a further representation that the conditions precedent to the Contractor's being entitled to final payment as set forth in Subparagraph 9.7.2 have been fulfilled. However, by issuing a Certificate for Payment, the Architect shall not thereby be deemed to represent that he has made exhaustive or continuous on-site inspections to check the quality or quantity of the Work or that he has reviewed the construction means, methods, techniques, sequences or procedures, or that he has made any examination to ascertain how or for what purpose the Contractor has used the moneys previously paid on account of the Contract Sum.

9.4.3 After the Architect has issued a Certificate for Payment, the Owner shall make payment in the manner provided in the Agreement.

9.4.4 No Certificate for a progress payment, nor any progress payment, nor any partial or entire use or occupancy of the Project by the Owner, shall constitute an acceptance of any Work not in accordance with the Contract Documents.

9.5 PAYMENTS WITHHELD
9.5.1 The Architect may decline to approve an Application for Payment and may withhold his Certificate in whole or in part if in his opinion he is unable to make representations to the Owner as provided in Subparagraph 9.4.2. The Architect may also decline to approve any Applications for Payment or, because of subsequently discovered evidence or subsequent inspections, he may nullify the whole or any part of any Certificate for Payment previously issued to such extent as may be necessary in his opinion to protect the Owner from loss because of:
    .1 defective work not remedied,
    .2 claims filed or reasonable evidence indicating probable filing of claims,
    .3 failure of the Contractor to make payments properly to Subcontractors or for labor, materials or equipment,   .4 reasonable doubt that the Work can be completed for the unpaid balance of the Contract Sum,
    .5 damage to another contractor,
    .6 reasonable indication that the Work will not be completed within the Contract Time, or
    .7 unsatisfactory prosecution of the Work by the Contractor.

9.5.2 When the above grounds in Subparagraph 9.5.1 are removed, payment shall be made for amounts withheld because of them.

9.6 FAILURE OF PAYMENT
9.6.1 If the Architect should fail to issue any Certificate for Payment, through no fault of the Contractor, within seven days after receipt of the Contractor's Application for Payment, or if the Owner should fail to pay the Contractor within seven days after the date of payment established in the Agreement any amount certified by the Architect or awarded by arbitration, then the Contractor may, upon seven additional days' written notice to the

Owner and the Architect, stop the Work until payment of the amount owing has been received.

9.7 SUBSTANTIAL COMPLETION AND FINAL PAYMENT

9.7.1   When the Contractor determines that the Work or a designated portion thereof acceptable to the Owner is substantially complete, the Contractor shall prepare for submission to the Architect a list of items to be completed or corrected. The failure to include any items on such list does not alter the responsibility of the Contractor to complete all Work in accordance with the Contract Documents. When the Architect on the basis of an inspection determines that the Work is substantially complete, he will then prepare a Certificate of Substantial Completion which shall establish the Date of Substantial Completion, shall state the responsibilities of the Owner and the Contractor for maintenance, heat, utilities, and insurance, and shall fix the time within which the Contractor shall complete the items listed therein, said time to be within the Contract Time unless extended pursuant to Paragraph 8.3. The Certificate of Substantial Completion shall be submitted to the Owner and the Contractor for their written acceptance of the responsibilities assigned to them in such Certificate.

9.7.2   Upon receipt of written notice that the Work is ready for final inspection and acceptance and upon receipt of a final Application for Payment, the Architect will promptly make such inspection and, when he finds the Work acceptable under the Contract Documents and the Contract fully performed, he will promptly issue a final Certificate for Payment stating that to the best of his knowledge, information and belief, and on the basis of his observations and inspections, the Work has been completed in accordance with the terms and conditions of the Contract Documents and that the entire balance found to be due the Contractor, and noted in said final Certificate, is due and payable.

9.7.3   Neither the final payment nor the remaining retained percentage shall become due until the Contractor submits to the Architect (1) an Affidavit that all payrolls, bills for materials and equipment, and other indebtedness connected with the Work for which the Owner or his property might in any way be responsible, have been paid or otherwise satisfied, (2) consent of surety, if any, to final payment and (3), if required by the Owner, other data establishing payment or satisfaction of all such obligations, such as receipts, releases and waivers of liens arising out of the Contract, to the extent and in such form as may be designated by the Owner. If any Subcontractor refuses to furnish a release or waiver required by the Owner, the Contractor may furnish a bond satisfactory to the Owner to indemnify him against any such lien. If any such lien remains unsatisfied after all payments are made, the Contractor shall refund to the Owner all moneys that the latter may be compelled to pay in discharging such lien, including all costs and reasonable attorneys' fees.

9.7.4   If after Substantial Completion of the Work final completion thereof is materially delayed through no fault of the Contractor, and the Architect so confirms, the Owner shall, upon certification by the Architect, and without terminating the Contract, make payment of the balance due for that portion of the Work fully completed and accepted. If the remaining balance for Work not fully completed or corrected is less than the retainage stipulated in the Agreement, and if bonds have been furnished as required in Subparagraph 7.5.1, the written consent of the surety to the payment of the balance due for that portion of the Work fully completed and accepted shall be submitted by the Contractor to the Architect prior to certification of such payment. Such payment shall be made under the terms and conditions governing final payment, except that it shall not constitute a waiver of claims.

9.7.5   The making of final payment shall constitute a waiver of all claims by the Owner except those arising from:

.1 unsettled liens,

.2 faulty or defective Work appearing after Substantial Completion,

.3 failure of the Work to comply with the requirements of the Contract Documents, or

.4 terms of any special guarantees required by the Contract Documents.

9.7.6   The acceptance of final payment shall constitute a waiver of all claims by the Contractor except those previously made in writing and still unsettled.

ARTICLE 10

PROTECTION OF PERSONS AND PROPERTY

10.1   SAFETY PRECAUTIONS AND PROGRAMS

10.1.1   The Contractor shall be responsible for initiating, maintaining and supervising all safety precautions and programs in connection with the Work.

10.2   SAFETY OF PERSONS AND PROPERTY

10.2.1   The Contractor shall take all reasonable precautions for the safety of, and shall provide all reasonable protection to prevent damage, injury or loss to:

.1 all employees on the Work and all other persons who may be affected thereby;

.2 all the Work and all materials and equipment to be incorporated therein, whether in storage on or off the site, under the care, custody or control of the Contractor or any of his Subcontractors or Sub-subcontractors; and

3. other property at the site or adjacent thereto, including trees, shrubs, lawns, walks, pavements, roadways, structures and utilities not designated for removal, relocation or replacement in the course of construction.

10.2.2   The Contractor shall comply with all applicable laws, ordinances, rules, regulations and orders of any public authority having jurisdiction for the safety of persons or property or to protect them from damage, injury or loss. He shall erect and maintain, as required by existing conditions and progress of the Work, all reasonable safeguards for safety and protection, including posting danger signs and other warnings against hazards, promulgating safety regulations and notifying owners and users of adjacent utilities.

10.2.3   When the use or storage of explosives or other hazardous materials or equipment is necessary for the execution of the Work, the Contractor shall exercise the utmost care and shall carry on such activities under the supervision of properly qualified personnel.

10.2.4   All damage or loss to any property referred to in Clauses 10.2.1.2 and 10.2.1.3 caused in whole or in part by the Contractor, any Subcontractor, any Sub-subcontractor, or anyone directly or indirectly employed by any of them, or by anyone for whose acts any of them may be liable, shall be remedied by the Contractor, except damage or loss attributable to faulty Drawings or Specifications or to the acts or omissions of the Owner or Architect or anyone employed by either of them or for whose acts either of them may be liable, and not attributable to the fault or negligence of the Contractor.

10.2.5   The Contractor shall designate a responsible member of his organization at the site whose duty shall be the prevention of accidents. This person shall be the Contractor's superintendent unless otherwise designated in writing by the Contractor to the Owner and the Architect.

10.2.6   The Contractor shall not load or permit any part of the Work to be loaded so as to endanger its safety.

10.3   EMERGENCIES

10.3.1   In any emergency affecting the safety of persons or property, the Contractor shall act, at his discretion, to prevent threatened damage, injury or loss. Any additional compensation or extension of time claimed by the Contractor on account of emergency work shall be determined as provided in Article 12 for Changes in the Work.

ARTICLE 11

INSURANCE

11.1   CONTRACTOR'S LIABILITY INSURANCE

11.1.1   The Contractor shall purchase and maintain such insurance as will protect him from claims set forth below which may arise out of or result from the Contractor's operations under the Contract, whether such operations be by himself or by any Subcontractor or by anyone directly or indirectly employed by any of them, or by anyone for whose acts any of them may be liable:

.1   claims under workmen's compensation, disability benefit and other similar employee benefit acts;

.2   claims for damages because of bodily injury, occupantional sickness or disease, or death of his employees, and claims insured by usual personal injury liability coverage;

.3   claims for damages because of bodily injury, sickness or disease, or death of any person other than his employees, and claims insured by usual personal injury liability coverage; and

.4   claims for damages because of injury to or destruction of tangible property, including loss of use resulting therefrom.

11.1.2   The insurance required by Subparagraph 11.1.1 shall be written for not less than any limits of liability specified in the Contract Documents, or required by law, whichever is greater, and shall include contractual liability insurance as applicable to the Contractor's obligations under Paragraph 4.18.

11.1.3   Certificates of Insurance acceptable to the Owner shall be filed with the Owner prior to commencement of the Work. These Certificates shall contain a provision that coverages afforded under the policies will not be cancelled until at least fifteen days' prior written notice has been given to the Owner.

11.2   OWNER'S LIABILITY INSURANCE

11.2.1   The Owner shall be responsible for purchasing and maintaining his own liability insurance and, at his option, may purchase and maintain such insurance as will protect him against claims which may arise from operations under the Contract.

11.3   PROPERTY INSURANCE

11.3.1   Unless otherwise provided, the Owner shall purchase and maintain property insurance upon the entire Work at the site to the full insurable value thereof. This insurance shall include the interests of the Owner, the Contractor, Subcontractors and Sub-subcontractors in the Work and shall insure against the perils of Fire, Extended Coverage, Vandalism and Malicious Mischief.

11.3.2   The Owner shall purchase and maintain such steam boiler and machinery insurance as may be required by the Contract Documents or by law. This insurance shall include the interests of the Owner, the Contractor, Subcontractors and Sub-subcontractors in the Work.

11.3.3   Any insured loss is to be adjusted with the Owner and made payable to the Owner as trustee for the insureds, as their interests may appear, subject to the requirements of any applicable mortgagee clause and of Subparagraph 11.3.8.

11.3.4   The Owner shall file a copy of all policies with the Contractor before an exposure to loss may occur. If the Owner does not intend to purchase such insurance, he shall inform the Contractor in writing prior to commencement of the Work. The Contractor may then effect insurance which will protect the interests of himself, his Subcontractors and the Sub-subcontractors in the Work, and by appropriate Change Order the cost thereof shall be charged to the Owner. If the Contractor is damaged by failure of the Owner to purchase or maintain such insurance and so to notify the Contractor, then the Owner shall bear all reasonable costs properly attributable thereto.

11.3.5   If the Contractor requests in writing that other special insurance be included in the property insurance policy, the Owner shall, if possible, include such insurance, and the cost thereof shall be charged to the Contractor by appropriate Change Order.

11.3.6   The Owner and Contractor waive all rights against each other for damages caused by fire or other perils to the extent covered by insurance provided under this Paragraph 11.3, except such rights as they may have to the proceeds of such insurance held by the Owner as trustee. The Contractor shall require similar waivers by Subcontractors and Sub-subcontractors in accordance with Clause 5.3.1.5.

11.3.7   If required in writing by any party in interest, the Owner as trustee shall, upon the occurrence of an insured loss, give bond for the proper performance of his duties. He shall deposit in a separate account any money so received, and he shall distribute it in accordance with such agreement as the parties in interest may reach, or in accordance with an award by arbitration in which case the procedure shall be as provided in Paragraph 7.10. If after such loss no other special agreement is made, replacement of damaged work shall be covered by an appropriate Change Order.

11.3.8   The Owner as trustee shall have power to adjust and settle any loss with the insurers unless one of the parties in interest shall object in writing within five days after the occurrence of loss to the Owner's exercise of this power, and if such objection be made, arbitrators shall be chosen as provided in Paragraph 7.10. The Owner as trustee shall, in that case, make settlement with the insurers in accordance with the directions of such arbitrators. If distribution of the insurance proceeds by arbitration is required, the arbitrators will direct such distribution.

11.4   LOSS OF USE INSURANCE
11.4.1   The Owner, at his option, may purchase and maintain such insurance as will insure him against loss of use of his property due to fire or other hazards, however caused.

ARTICLE 12

CHANGES IN THE WORK

12.1   CHANGE ORDERS

12.1.1   The Owner, without invalidating the Contract, may order Changes in the Work within the general scope of the Contract consisting of additions, deletions or other revisions, the Contract Sum and the Contract Time being adjusted accordingly. All such Changes in the Work shall be authorized by Change Order, and shall be executed under the applicable conditions of the Contract Documents.

12.1.2   A Change Order is a written order to the Contractor signed by the Owner and the Architect, issued after the execution of the Contract, authorizing a Change in the Work or an adjustment in the Contract Sum or the Contract Time. Alternatively, the Change Order may be signed by the Architect alone, provided he has written authority from the Owner for such procedure and that a copy of such written authority is furnished to the Contractor upon request. The Contract Sum and the Contract Time may be changed only by Change Order.

12.1.3   The cost or credit to the Owner resulting from a Change in the Work shall be determined in one or more of the following ways:
    .1 by mutual acceptance of a lump sum properly itemized;
    .2 by unit prices stated in the Contract Documents or subsequently agreed upon; or
    .3 by cost and a mutually acceptable fixed or percentage fee.

12.1.4   If none of the methods set forth in Subparagraph 12.1.3 is agreed upon, the Contractor, provided he receives a Change Order, shall promptly proceed with the Work involved. The cost of such Work shall then be determined by the Architect on the basis of the Contractor's reasonable expenditures and savings, including, in the case of an increase in the Contract Sum, a reasonable allowance for overhead and profit. In such case, and also under Clause 12.1.3.3 above, the Contractor shall keep and present, in such form as the Architect may prescribe, an itemized accounting together with appropriate supporting data. Pending final determination of cost to the Owner, payments on account shall be made on the Architect's Certificate for Payment. The amount of credit to be allowed by the Contractor to the Owner for any deletion or change which results in a net decrease in cost will be the amount of the actual net decrease as confirmed by the Architect. When both additions and credits are involved in any one change, the allowance for overhead and profit shall be figured on the basis of net increase, if any.

12.1.5   If unit prices are stated in the Contract Documents or subsequently agreed upon, and if the quantities originally contemplated are so changed in a proposed Change Order that application of the agreed unit prices to the quantities of Work proposed will create a hardship on the Owner or the Contractor, the applicable unit prices shall be equitably adjusted to prevent such hardship.

12.1.6   Should concealed conditions encountered in the performance of the Work below the surface of the ground be at variance with the conditions indicated by the Contract Documents or should unknown physical conditions below the surface of the ground of an unusual nature, differing materially from those ordinarily encountered and generally recognized as inherent in work of the character provided for in this Contract, be encountered, the Contract Sum shall be equitably adjusted by Change Order upon claim by either party made within a reasonable time after the first observance of the conditions.

12.1.7   If the Contractor claims that additional cost or time is involved because of (1) any written interpretation issued pursuant to Subparagraph 1.2.5, (2) any order by the Architect to stop the Work pursuant to Subparagraph 2.2.12 where the Contractor was not at fault, or (3) any written order for a minor change in the Work issued pursuant to Paragraph 12.3, the Contractor shall make such claim as provided in Paragraph 12.2.

12.2   CLAIMS FOR ADDITIONAL COST OR TIME

12.2.1   If the Contractor wishes to make a claim for an increase in the Contract Sum or an extension in the Contract Time, he shall give the Architect written notice thereof within a reasonable time after the occurrence of the event giving rise to such claim. This notice shall be given by the Contractor before proceeding to execute the Work, except in an emergency endangering life or property in which case the Contractor shall proceed in accordance with Subparagraph 10.3.1. No such claim shall be valid unless so made. If the Owner and the Contractor cannot agree on the amount of the adjustment in the Contract Sum or the Contract Time, it shall be determined by the Architect. Any change in the Contract Sum or Contract Time resulting from such claim shall be authorized by Change Order.

12.3   MINOR CHANGES IN THE WORK

12.3.1   The Architect shall have authority to order minor changes in the Work not involving an adjustment in the Contract Sum or an extension of the Contract Time and not inconsistent with the intent of the Contract Documents. Such changes may be effected by Field Order or by other written order. Such changes shall be binding on the Owner and the Contractor.

12.4   FIELD ORDERS

12.4.1   The Architect may issue written Field Orders which interpret the Contract Documents in accordance with Subparagraph 1.2.5 or which order minor changes in the Work in accordance with Paragraph 12.3 without change in Contract Sum or Contract Time. The Contractor shall carry out such Field Orders promptly.

ARTICLE 13

UNCOVERING AND CORRECTION OF WORK

13.1   UNCOVERING OF WORK

13.1.1   If any Work should be covered contrary to the request of the Architect, it must, if required by the Architect, be uncovered for his observation and replaced at the Contractor's expense.

13.1.2   If any other Work has been covered which the Architect has not specifically requested to observe prior to being covered, the Architect may request to see such Work and it shall be uncovered by the Contractor. If such Work be found in accordance with the Contract Documents, the cost of uncovering and replacement shall, by appropriate Change Order, be charged to the Owner. If such Work be found not in accordance with the Contract Documents, the Contractor shall pay such costs unless it be found that this condition was caused by a separate contractor employed as provided in Article 6, and in that event the Owner shall be responsible for the payment of such costs.

13.2   CORRECTION OF WORK

13.2.1   The Contractor shall promptly correct all Work rejected by the Architect as defective or as failing to conform to the Contract Documents whether observed before or after Substantial Completion and whether or not fabricated, installed or completed. The Contractor shall bear all costs of correcting such rejected Work, including the cost of the Architect's additional services thereby made necessary.

13.2.2   If, within one year after the Date of Substantial Completion or within such longer period of time as may be prescribed by law or by the terms of any applicable special guarantee required by the Contract Documents, any of the Work is found to be defective or not in accordance with the Contract Documents, the Contractor shall correct it promptly after receipt of a written notice from the Owner to do so unless the Owner has previously given the Contractor a written acceptance of such condition. The Owner shall give such notice promptly after discovery of the condition.

13.2.3   All such defective or non-conforming Work under Subparagraphs 13.2.1 and 13.2.2 shall be removed from the site where necessary, and the Work shall be corrected to comply with the Contract Documents without cost to the Owner.

13.2.4   The Contractor shall bear the cost of making good all work of separate contractors destroyed or damaged by such removal or correction.

13.2.5   If the Contractor does not remove such defective or non-conforming Work within a reasonable time fixed by written notice from the Architect, the Owner may remove it and may store the materials or equipment at the expense of the Contractor. If the Contractor does not pay the cost of such removal and storage within ten days thereafter, the Owner may upon ten additional days' written notice sell such Work at auction or at private sale and shall account for the net proceeds thereof, after deducting all the costs that should have been borne by the Contractor including compensation for additonal architectural services. If such proceeds of sale do not cover all costs which the Contractor should have borne, the difference shall be charged to the Contractor and an appropriate Change Order shall be issued. If the payments then or thereafter due the Contractor are not sufficient to cover such amount, the Contractor shall pay the difference to the Owner.

13.2.6   If the Contractor fails to correct such defective or non-conforming Work, the Owner may correct it in accordance with Paragraph 7.6.

13.2.7   The obligations of the Contractor under this Paragraph 13.2 shall be in addition to and not in limitation of any obligations imposed upon him by special guarantees required by the Contract Documents or otherwise prescribed by law.

13.3   ACCEPTANCE OF DEFECTIVE OR NON-CONFORMING WORK
13.3.1   If the Owner prefers to accept defective or non-conforming Work, he may do so instead of requiring its removal and correction, in which case a Change Order will be issued to reflect an appropriate reduction in the Contract Sum, or, if the amount is determined after final payment, it shall be paid by the Contractor.

ARTICLE 14

TERMINATION OF THE CONTRACT

14.1   TERMINATION BY THE CONTRACTOR
14.1.1   If the Work is stopped for a period of thirty days under an order of any court or other public authority having jurisdiction, through no act or fault of the Contractor or a

Subcontractor or their agents or employees or any other persons performing any of the Work under a contract with the Contractor, or if the Work should be stopped for a period of thirty days by the Contractor for the Architect's failure to issue a Certificate for Payment as provided in Paragraph 9.6 or for the Owner's failure to make payment thereon as provided in Paragraph 9.6, then the Contractor may, upon seven days' written notice to the Owner and the Architect, terminate the Contract and recover from the Owner payment for all Work executed and for any proven loss sustained upon any materials, equipment, tools, construction equipment and machinery, including reasonable profit and damages.

14.2 TERMINATION BY THE OWNER

14.2.1 If the Contractor is adjudged a bankrupt, or if he makes a general assignment for the benefit of his creditors, or if a receiver is appointed on account of his insolvency, or if he persistently or repeatedly refuses or fails, except in cases for which extension of time is provided, to supply enough properly skilled workmen or proper materials, or if he fails to make prompt payment to Subcontractors or for materials or labor, or persistently disregards laws, ordinances, rules, regulations or orders of any public authority having jurisdiction, or otherwise is guilty of a substantial violation of a provision of the Contract Documents, then the Owner, upon certification by the Architect that sufficient cause exists to justify such action, may, without prejudice to any right or remedy and after giving the Contractor and his surety, if any, seven days' written notice, terminate the employment of the Contractor and take possession of the site and of all materials, equipment, tools, construction equipment and machinery thereon owned by the Contractor and may finish the Work by whatever method he may deem expedient. In such case the Contractor shall not be entitled to receive any further payment until the Work is finished.

14.2.2 If the unpaid balance of the Contract Sum exceeds the costs of finishing the Work, including compensation for the Architect's additional services, such excess shall be paid to the Contractor. If such costs exceed such unpaid balance, the Contractor shall pay the difference to the Owner. The costs incurred by the Owner as herein provided shall be certified by the Architect.

Standard Form 23-A, June 1964 Edition
General Services Administration, Fed. Proc. Reg. (41 CPR) 1-16.401

GENERAL PROVISIONS
(Construction Contract)

1. DEFINITIONS

(a) The term "head of the agency" or "Secretary" as used herein means the Secretary, the Under Secretary, any Assistant Secretary, or any other head or assistant head of the executive or military department or other Federal agency; and the term "his duly authorized representative" means any person or persons or board (other than the Contracting Officer) authorized to act for the head of the agency or the Secretary.

(b) The term "Contracting Officer" as used herein means the person executing this contract on behalf of the Government and includes a duly appointed successor or authorized representative.

2. SPECIFICATIONS AND DRAWINGS

The Contractor shall keep on the work a copy of the drawings and specifications and shall at all times give the Contracting Officer access thereto. Anything mentioned in the specifications and not shown on the drawings, or shown on the drawings and not mentioned in the specifications, shall be of like effect as if shown or mentioned in both. In case of difference between drawings and specifications, the specifications shall govern. In case of discrepancy either in the figures, in the drawings, or in the specifications, the matter shall be promptly submitted to the Contracting Officer, who shall promptly make a determination in writing. Any adjustment by the Contractor without such a determination shall be at his own risk and expense. The Contracting Officer shall furnish from time to time such detail drawings and other information as he may consider necessary, unless otherwise provided.

3. CHANGES

The Contracting Officer may, at any time, by written order, and without notice to the sureties, make changes in the drawings and/or specifications of this contract if within its general scope. If such changes cause an increase or decrease in the Contractor's cost of, or time required for, performance of the contract, an equitable adjustment shall be made and the contract modified in writing accordingly. Any claim of the Contractor for adjustment under this clause must be asserted in writing within 30 days from the date of receipt by the Contractor of the notification of change unless the Contracting Officer grants a further period of time before the date of final payment under the contract. If the parties fail to agree upon the adjustment to be made, the dispute shall be determined as provided in Clause 6 of these General Provisions; but nothing provided in this clause shall excuse the Contractor from proceeding with the prosecution of the work as changed. Except as otherwise provided in this contract, no charge for any extra work or material will be allowed.

4. CHANGED CONDITIONS

The Contractor shall promptly, and before such conditions are disturbed, notify the Contracting Officer in writing of: (a) subsurface or latent physical conditions at the site differing materially from those indicated in this contract, or (b) unknown physical conditions at the site, of an unusual nature, differing materially from those ordinarily encountered and generally recognized as inhering in work of the character provided for in this contract. The Contracting Officer shall promptly investigate the conditions, and if he finds that such conditions do so materially differ and cause an increase or decrease in the Contractor's cost of, or the time required for, performance of this contract, an equitable adjustment shall be made and the contract modified in writing accordingly. Any claim of the Contractor for adjustment hereunder shall not be allowed unless he has given notice as above required; or unless the Contracting Officer grants a further period of time before the date of final payment under the contract. If the parties fail to agree upon the adjustment to be made, the dispute shall be determined as provided in Clause 6 of these General Provisions.

5. TERMINATION FOR DEFAULT–DAMAGES FOR DELAY–TIME EXTENSIONS

(a) If the Contractor refuses or fails to prosecute the work, or any separable part thereof, with such diligence as will insure its completion within the time specified in this contract, or any extension thereof, or fails to complete said work within such time, the Government may, by written notice to the Contractor, terminate his right to proceed with the work or such part of the work as to which there has been delay. In such event the Government may take over the work and prosecute the same to completion, by contract or otherwise, and may take possession of and utilize in completing the work such materials, appliances, and plant as may be on the site of the work and necessary therefor. Whether or not the Contractor's right to proceed with the work is terminated, he and his sureties shall be liable for any damage to the Government resulting from his refusal or failure to complete the work within the specified time.

(b) If fixed and agreed liquidated damages are provided in the contract and if the Government so terminates the Contractor's right to proceed, the resulting damage will consist of such liquidated damages until such reasonable time as may be required for final completion of the work together with any increased costs occasioned the Government in completing the work.

(c) If fixed and agreed liquidated damages are provided in the contract and if the Government does not so terminate the Contractor's right to proceed, the resulting damage will consist of such liquidated damages until the work is completed or accepted.

(d) The Contractor's right to proceed shall not be so terminated nor the Contractor charged with resulting damage if:

(1) The delay in the completion of the work arises from unforeseeable causes beyond the control and without the fault or negligence of the Contractor, including but not restricted to, acts of God, acts of the public enemy, acts of the Government in either its sovereign or contractual capacity, acts of another contractor in the performance of a contract with the Government, fires, floods, epidemics, quarantine restrictions, strikes, freight embargoes, unusually severe weather, or delays of subcontractors or suppliers arising from unforeseeable causes beyond the control and without the fault or negligence of both the Contractor and such subcontractors or suppliers; and

(2) The Contractor, within 10 days from the beginning of any such delay (unless the Contracting Officer grants a further period of time before the date of final payment under the contract), notifies the Contracting Officer in writing of the causes of delay.

The Contracting Officer shall ascertain the facts and the extent of the delay and extend the time for completing the work when, in his judgment, the findings of fact justify such an extension, and his findings of fact shall be final and conclusive on the parties, subject only to appeal as provided in Clause 6 of these General Provisions.

(e) If, after notice of termination of the Contractor's right to proceed under the provisions of this clause, it is determined for any reason that the Contractor was not in default under the provisions of this clause, or that the delay was excusable under the provisions of this clause, the rights and obligations of the parties shall, if the contract contains a clause providing for termination for convenience of the Government, be the same as if the notice of termination had been issued pursuant to such clause. If, in the foregoing circumstances, this contract does not contain a clause providing for termination for convenience of the Government, the contract shall be equitably adjusted to compensate for such termination and the contract modified accordingly; failure to agree to any such adjustment shall be a dispute concerning a question of fact within the meaning of the clause of this contract entitled "Disputes."

(f) The rights and remedies of the Government provided in this clause are in addition to any other rights and remedies provided by law or under this contract.

6. DISPUTES

(a) Except as otherwise provided in this contract, any dispute concerning a question of fact arising under this contract which is not disposed of by agreement shall be decided by the Contracting Officer, who shall reduce his decision to writing and mail or otherwise furnish a copy thereof to the Contractor. The decision of the Contracting Officer shall be final and conclusive unless, within 30 days from the date of receipt of such copy, the Contractor mails or otherwise furnishes to the Contracting Officer a written appeal addressed to the head of the agency involved. The decision of the head of the agency or his duly authorized representative for the determination of such appeals shall be final and conclusive. This provision shall not be pleaded in any suit involving a question of fact arising under this contract as limiting judicial review of any such decision to cases where fraud by such official or his representative or board is alleged: *Provided, however,* that any such decision shall be final and conclusive unless the same is fraudulent or capricious or arbitrary or so grossly erroneous as necessarily to imply bad faith or is not supported by substantial evidence. In connection with any appeal proceeding under this clause, the Contractor shall be afforded an opportunity to be heard and to offer evidence in support of his appeal. Pending final decision of a dispute hereunder, the Contractor shall proceed diligently with the performance of the contract and in accordance with the Contracting Officer's decision.

(b) This Disputes clause does not preclude consideration of questions of law in connection with decisions provided for in paragraph (a) above. Nothing in this contract, however, shall be construed as making final the decision of any administrative official, representative, or board on a question of law.

7. PAYMENTS TO CONTRACTOR

(a) The Government will pay the contract price as hereinafter provided.

(b) The Government will make progress payments monthly as the work proceeds, or at more frequent intervals as determined by the Contracting Officer, on estimates approved by the Contracting Officer. If requested by the Contracting Officer, the Contractor shall furnish a breakdown of the total contract price showing the amount included therein for each principal category of the work, in such detail as requested, to provide a basis for determining progress payments. In the preparation of estimates the Contracting Officer, at his discretion, may authorize material delivered on the site and preparatory work done to be taken into consideration. Material delivered to the Contractor at locations other than the site may also be taken into consideration (1) if such consideration is specifically authorized by the contract and (2) if the Contractor furnishes satisfactory evidence that he has acquired title to such material and that it will be utilized on the work covered by this contract.

(c) In making such progress payments, there shall be retained 10 percent of the estimated amount until final completion and acceptance of the contract work. However, if the Contracting Officer, at any time after 50 percent of the work has been completed, finds that satisfactory progress is being made, he may authorize any of the remaining progress payments to be made in full. Also, whenever the work is substantially complete, the Contracting Officer, if he considers the amount retained to be in excess of the amount adequate for the protection of the Government, at his discretion, may release to the Contractor all or a portion of such excess amount. Furthermore, on completion and acceptance of each separate building, public work, or other division of the contract, on which the price is stated separately in the contract, payment may be made therefor without retention of a percentage.

(d) All material and work covered by progress payments made shall thereupon become the sole property of the Government, but this provision shall not be construed as relieving the Contractor from the sole responsibility for all material and work upon which payments have been made or the restoration of any damaged work, or as waiving the right of the Government to require the fulfillment of all of the terms of the contract.

(e) Upon completion and acceptance of all work, the amount due the Contractor under this contract shall be paid upon the presentation of a properly executed voucher and after the Contractor shall have furnished the Government with a release, if required, of all claims against the Government arising by virtue of this contract, other than claims in stated amounts as may be specifically excepted by the Contractor from the operation of the release. If the Contractor's claim to amounts payable under the contract has been assigned under the Assignment of Claims Act of 1940, as amended (31 U.S.C. 203, 41 U.S.C. 15), a release may also be required of the assignee.

8. ASSIGNMENT OF CLAIMS

(a) Pursuant to the provisions of the Assignment of Claims Act of 1940, as amended (31 U.S.C. 203, 41 U.S.C. 15), if this contract provides for payments aggregating $1,000 or more, claims for moneys due or to become due the Contractor from the Government under this contract may be assigned to a bank, trust company, or other financing institution, including any Federal lending agency, and may thereafter be further assigned and reassigned to any such institution. Any such assignment or reassignment shall cover all amounts payable under this contract and not already paid, and shall not be made to more than one party, except that any such assignment or reassignment may be made to one party as agent or trustee for two or more parties participating in such financing. Unless otherwise provided in this contract, payments to an assignee of any moneys due or to become due under this contract shall not, to the extent provided in said Act, as amended, be subject to reduction or setoff. (The preceding sentence applies only if this contract is made in time of war or national emergency as defined in said Act; and is with the Department of Defense, the General Services Administration, the Atomic Energy Commission, the National Aeronautics and Space Administration, the Federal Aviation Agency, or any other department or agency of the United States designated by the President pursuant to Clause 4 of the proviso of section 1 of the Assignment of Claims Act of 1940, as amended by the Act of May 15, 1951, 65 Stat. 41.)

(b) In no event shall copies of this contract or of any plans, specifications, or other similar documents relating to work under this contract, if marked "Top Secret," "Secret," or "Confidential," be furnished to any assignee of any claim arising under this contract or to any other person not entitled to receive the same. However, a copy of any part or all of this contract so marked may be furnished, or any information contained therein may be disclosed, to such assignee upon the prior written authorization of the Contracting Officer.

9. MATERIAL AND WORKMANSHIP

(a) Unless otherwise specifically provided in this contract, all equipment, material, and articles incorporated in the work covered by this contract are to be new and of the most suitable grade for the purpose intended. Unless otherwise specifically provided in this contract, reference to any equipment, material, article, or patented process, by trade name, make, or catalog number, shall be regarded as establishing a standard of quality and shall not be construed as limiting competition, and the Contractor may, at his option, use any equipment, material, article, or process which, in the judgment of the Contracting Officer, is equal to that named. The Contractor shall furnish to the Contracting Officer for his approval the name of the manufacturer, the model number, and other identifying data and information respecting the performance, capacity, nature, and rating of the machinery and mechanical and other equipment which the Contractor contemplates incorporating in the work. When required by this contract or when called for by the Contracting Officer, the Contractor shall furnish the Contracting Officer for approval full information concerning the material or articles which he contemplates incorporating in the work. When so directed, samples shall be submitted for approval at the Contractor's expense, with all shipping charges prepaid. Machinery, equipment, material, and articles installed or used without required approval shall be at the risk of subsequent rejection.

(b) All work under this contract shall be performed in a skillful and workmanlike manner. The Contracting Officer may, in writing, require the Contractor to remove from the work any employee the Contracting Officer deems incompetent, careless, or otherwise objectionable.

10. INSPECTION AND ACCEPTANCE

(a) Except as otherwise provided in this contract, inspection and test by the Government of material and workmanship required by this contract shall be made at reasonable times and at the site of the work, unless the Contracting Officer determines that such inspection or test of material which is to be incorporated in the work shall be made at the place of production, manufacture, or shipment of such material. To the extent specified by the Contracting Officer at the time of determining to make off-site inspection or test, such inspection or test shall be conclusive as to whether the material involved conforms to the contract requirements. Such offsite inspection or test shall not relieve the Contractor of responsibility for damage to or loss of the material prior to acceptance, nor in any way affect the continuing rights of the Government after acceptance of the completed work under the terms of paragraph (f) of this clause, except as hereinabove provided.

(b) The Contractor shall, without charge, replace any material or correct any workmanship found by the Government not to conform to the contract requirements, unless in the public interest the Government consents to accept such material or workmanship with an appropriate adjustment in contract price. The Contractor shall promptly segregate and remove rejected material from the premises.

(c) If the Contractor does not promptly replace rejected material or correct rejected workmanship, the Government (1) may, by contract or otherwise, replace such material or correct such workmanship and charge the cost thereof to the Contractor, or (2) may terminate the Contractor's right to proceed in accordance with Clause 5 of these General Provisions.

(d) The Contractor shall furnish promptly, without additional charge, all facilities, labor, and material reasonably needed for performing such safe and convenient inspection and test as may be required by the Contracting Officer. All inspection and test by the Government shall be performed in such manner as not unnecessarily to delay the work. Special, full size, and performance tests shall be performed as described in this contract. The Contractor shall be charged with any additional cost of inspection when material and workmanship are not ready at the time specified by the Contractor for its inspection.

(e) Should it be considered necessary or advisable by the Government at any time before acceptance of the entire work to make an examination of work already completed, by removing or tearing out same, the Contractor shall, on request, promptly furnish all necessary facilities, labor, and material. If such work is found to be defective or nonconforming in any material respect, due to the fault of the Contractor or his subcontractors, he shall defray all the expenses of such examination and of satisfactory reconstruction. If, however, such work is found to meet the requirements of the contract, an equitable adjustment shall be made in the contract price to compensate the Contractor for the additional services involved in such examination and reconstruction and, if completion of the work has been delayed thereby, he shall, in addition, be granted a suitable extension of time.

(f) Unless otherwise provided in this contract, acceptance by the Government shall be made as promptly as practicable after completion and inspection of all work required by this contract. Acceptance shall be final and conclusive except as regards latent defects, fraud, or such gross mistakes as may amount to fraud, or as regards the Government's rights under any warranty or guarantee.

11. SUPERINTENDENCE BY CONTRACTOR

The Contractor shall give his personal superintendence to the work or have a competent foreman or superintendent, satisfactory to the Contracting Officer, on the work at all times during progress, with authority to act for him.

12. PERMITS AND RESPONSIBILITIES

The Contractor shall, without additional expense to the Government, be responsible for obtaining any necessary licenses and permits, and for complying with any applicable Federal, State, and municipal laws, codes, and regulations, in connection with the prosecution of the work. He shall be similarly responsible for all damages to persons or property that occur as a result of his fault or negligence. He shall take proper safety and health precautions to protect the work, the workers, the public, and the property of others. He shall also be responsible for all materials delivered and work performed until completion and acceptance of the entire construction work, except for any completed unit of construction thereof which theretofore may have been accepted.

13. CONDITIONS AFFECTING THE WORK

The Contractor shall be responsible for having taken steps reasonably necessary to ascertain the nature and location of the work, and the general and local conditions which can affect the work or the cost thereof. Any failure by the Contractor to do so will not relieve him from responsibility for successfully performing the work without additional expense to the Government. The Government assumes no responsibility for any understanding or representations concerning conditions made by any of its officers or agents prior to the execution of this contract, unless such understanding or representations by the Government are expressly stated in the contract.

14. OTHER CONTRACTS

The Government may undertake or award other contracts for additional work, and the Contractor shall fully cooperate with such other contractors and Government employees and carefully fit his own work to such additional work as may be directed by the Contracting Officer. The Contractor shall not commit or permit any act which will interfere with the performance of work by any other contractor or by Government employees.

15. PATENT INDEMNITY

Except as otherwise provided, the Contractor agrees to indemnify the Government and its officers, agents, and employees against liability, including costs and expenses, for infringement upon any Letters Patent of the United States (except Letters Patent issued upon an application which is now or may hereafter be, for reasons of national security, ordered by the Government to be kept secret or otherwise withheld from issue) arising out of the performance of this contract or out of the use or disposal by or for the account of the Government of supplies furnished or construction work performed hereunder.

16. ADDITIONAL BOND SECURITY

If any surety upon any bond furnished in connection with this contract becomes unacceptable to the Government, or if any such surety fails to furnish reports as to his financial condition from time to time as requested by the Government, the Contractor shall promptly furnish such additional security as may be required from time to time to protect the interests of the Government and of persons supplying labor or materials in the prosecution of the work contemplated by this contract.

17. COVENANT AGAINST CONTINGENT FEES

The Contractor warrants that no person or selling agency has been employed or retained to solicit or secure this contract upon an agreement or understanding for a commission, percentage, brokerage, or contingent fee, excepting bona fide employees or bona fide established commercial or selling agencies maintained by the Contractor for the purpose of securing business. For breach or violation of this warranty the Government shall have the right to annul this contract without liability or in its discretion to deduct from the contract price or consideration, or otherwise recover, the full amount of such commission, percentage, brokerage, or contingent fee.

18. OFFICIALS NOT TO BENEFIT

No member of Congress or resident Commissioner shall be admitted to any share or part of this contract, or to any benefit that may arise therefrom; but this provision shall not be construed to extend to this contract if made with a corporation for its general benefit.

19. BUY AMERICAN

(a) Agreement. In accordance with the Buy American Act (41 U.S.C. 10a–10d) and Executive Order 10582, December 17, 1954 (3 CFR Supp.), the Contractor agrees that only domestic construction material will be used (by the Contractor, subcontractors, materialmen, and suppliers) in the performance of this contract, except for nondomestic material listed in the contract.

(b) Domestic construction material. "Construction material" means any article, material, or supply brought to the construction site for incorporation in the building or work. An unmanufactured construction material is a "domestic construction material" if it has been mined or produced in the United States. A manufactured construction material is a "domestic construction material" if it has been manufactured in the United States and if the cost of its components which have been mined, produced, or manufactured in the United States exceeds 50 percent of the cost of all its components. "Component" means any article, material, or supply directly incorporated in a construction material.

(c) Domestic component. A component shall be considered to have been "mined, produced, or manufactured in the United States" (regardless of its source in fact) if the article, material, or supply in which it is incorporated was manufactured in the United States and the component is of a class or kind determined by the Government to be not mined, produced, or manufactured in the United States in sufficient and reasonably available commercial quantities and of a satisfactory quality.

20. CONVICT LABOR

In connection with the performance of work under this contract, the Contractor agrees not to employ any person undergoing sentence of imprisonment at hard labor.

21. EQUAL OPPORTUNITY CLAUSE

(The following clause is applicable unless this contract is exempt under the rules and regulations of the President's Committee on Equal Employment Opportunity (41 CFR, Chapter 60). Exemptions include contracts and subcontracts (i) not exceeding $10,000, (ii) not exceeding $100,000 for standard commercial supplies or raw materials, and (iii) under which work is performed outside the United States and no recruitment of workers within the United States is involved.)

During the performance of this contract, the Contractor agrees as follows:

(a) The Contractor will not discriminate against any employee or applicant for employment because of race, creed, color, or national origin. The Contractor will take affirmative action to ensure that applicants are employed, and that employees are treated during employment, without regard to their race, creed, color, or national origin. Such action shall include, but not be limited to, the following: employment, upgrading, demotion or transfer; recruitment or recruitment advertising; layoff or termination; rates of pay or other forms of compensation; and selection for training, including apprenticeship. The Contractor agrees to post in conspicuous places, available to employees and applicants for employment, notices to be provided by the Contracting Officer setting forth the provisions of this nondiscrimination clause.

(b) The Contractor will, in all solicitations or advertisements for employees placed by or on behalf of the Contractor, state that all qualified applicants will receive consideration for employment without regard to race, creed, color, or national origin.

(c) The Contractor will send to each labor union or representative of workers with which he has a collective bargaining agreement or other contract or understanding, a notice, to be provided by the agency Contracting Officer, advising the said labor union or workers' representative of the Contractor's commitments under this nondiscrimination clause, and shall post copies of the notice in conspicuous places available to employees and applicants for employment.

(d) The Contractor will comply with all provisions of Executive Order No. 10925 of March 6, 1961, as amended, and of the rules, regulations, and relevant orders of the President's Committee on Equal Employment Opportunity created thereby.

(e) The Contractor will furnish all information and reports required by Executive Order No. 10925 of March 6, 1961, as amended, and by the rules, regulations, and orders of the said Committee, or pursuant thereto, and will permit access to his books, records, and accounts by the contracting agency and the Committee for purposes of investigation to ascertain compliance with such rules, regulations, and orders.

(f) In the event of the Contractor's noncompliance with the nondiscrimination clause of this contract or with any of the said rules, regulations, or orders, this contract may be canceled, terminated, or suspended in whole or in part and the Contractor may be declared ineligible for further Government contracts in accordance with procedures authorized in Executive Order No. 10925 of March 6, 1961, as amended, and such other sanctions may be imposed and remedies invoked as provided in the said Executive order or by rule, regulation, or order of the President's Committee on Equal Employment Opportunity, or as otherwise provided by law.

(g) The Contractor will include the provisions of paragraphs (a) through (g) in every subcontract or purchase order unless exempted by rules, regulations, or orders of the President's Committee on Equal Employment Opportunity issued pursuant to section 303 of Executive Order No. 10925 of March 6, 1962, as amended, so that such providions will be binding upon each subcontractor or vendor. *The Contractor will take such action with respect to any subcontract or purchase order as the contracting agency may direct as a means of enforcing such provisions, including sanctions for noncompliance: Provided, however, that in the event the Contractor becomes involved in, or is threatened with, litigation with a subcontractor or vendor as a result of such direction by the contracting agency, the Contractor may request the United States to enter into such litigation to protect the interests of the United States.

*Unless otherwise provided, the Equal Opportunity Clause is not required to be inserted in subcontracts below the second tier except for subcontracts involving the performance of 'construction work' at the 'site of construction' (as those terms are defined in the Committee's rules and regulations) in which case the clause must be inserted in all such subcontracts. Subcontracts may incorporate by reference the Equal Opportunity Clause.

22. UTILIZATION OF SMALL BUSINESS CONCERNS

(a) It is the policy of the Government as declared by the Congress that a fair proportion of the purchases and contracts for supplies and services for the Government be placed with small business concerns.

(b) The Contractor agrees to accomplish the maximum amount of subcontracting to small business concerns that the Contractor finds to be consistent with the efficient performance of this contract.

Chapter 5

The Specification as A Contract Document

Nature of the Specifications[1]

By custom of long standing in the building industry documents pertaining to bidding requirements, sample contract forms, the General Conditions, and the Supplementary Conditions have been bound in a single volume together with the project Specification. This volume has commonly been called "The Specifications" and has thereby given rise to confusion between it and that element of the Contract Documents identified as the Specifications. In an effort to eliminate this confusion and maintain the necessary distinctions, the American Institute of Architects recommends the title "Project Manual" as better describing the volume's scope and content. Some A & E firms follow this concept, but use the title "Construction Documents." In both cases the use of these designations permits a systematic arrangement of requirements that are broken down into those pertaining to activities prior to execution of the construction contract and those that subsequently become a part of the contract. This procedure also underscores the point that the General Conditions and Supplementary Conditions are not specifications but contractual-legal provisions constituting separate elements of the Contract Documents. A typical table of contents for the Project Manual or Construction Documents Book might then show the following major divisions:

---

[1] There is considerable variation in usage between the singular and plural forms of this word without consistent difference in meaning. A project Specification is by its nature made up of many individual specifications covering separate segments or components of the work, and this probably accounts for its also being called the project Specifications. Both forms will be used interchangeably here (capitalized when referring to the project set as a whole).

Bidding Requirements
  Invitation to Bid (or Advertisement)
  Instructions to Bidders
  Proposal Form
Contract and Bond Forms
Conditions of the Contract
  General Conditions
  Supplementary Conditions
Specifications

The Conditions of the Contract were discussed in Chapter 4. Bidding requirements are treated in Chapter 7 and Agreements in Chapters 8 and 9.

As stated earlier the Specifications provide technical information concerning the building materials, components, systems and equipment indicated on the Drawings with respect to quality, performance characteristics, and results to be achieved by application of construction methods. The Drawings, on the other hand, portray graphically the work to be constructed with respect to dimensions, location, arrangement of components, materials, mechanical and electrical systems, and siting. Obviously, the functions of the Specifications and Drawings are complementary and it is important that their purposes and functions not be duplicated or interchanged. In order that the Specification may effectively perform its mission, careful consideration should be given to terms used in notes on the Drawings for identifying materials and equipment; these should be in general terms only so that more precise identification as to type, quality, composition, etc., may be controlled in the Specification.

Although it has become common practice to state in the General Conditions that all of the Contract Documents are complementary and that there is no intended order of precedence among them, there actually is a *de facto* precedence of the Specification over the Drawings. This is because the Specification is a written document that is more readily understood by persons outside the building industry (including the courts) than are graphic types of information. It is probable that the relative importance of the Specification in determining the intent of the Contract Documents will increase whether disputes are settled by litigation or arbitration.

Organization: The Uniform System

Traditionally the organization of the Specifications has been a matter of individual preference by the architect-engineer firm producing them. This naturally resulted in wide diversity of method, not only between one design office and another but between practices in different parts of the country. As the number of design firms and building contractors with nation-wide operations increased and nation-wide marketing of building products developed, a pressing need grew for a consistent arrangement of building construction specifications.

The first major response to this situation was made by the Construction Specifications Institute with publication in 1963 of the *CSI Format for Building Specifications.* This document promulgated the division-section concept of specification organization, establishing 16 basic groupings or divisions, each based on an interrelationship of the four factors of materials, trades, functions of work, and location of specified work. In the CSI Format, the word *section* denotes a basic unit of work, that is, an entity that generally describes a particular product or material and its installation. The term *division* therefore denotes a group of related sections. "The concept of grouping related sections into divisions was conceived as a practical means of providing national uniformity without disrupting current practices in specification writing. It provides a framework for specification arrangement without conflicting with normal bidding procedures or local trade jurisdictions."[2] There has been widespread acceptance of the division-section principle advocated by CSI within both federal and state agencies throughout the United States and Canada and among architects and engineers in private practice.

THE UNIFORM SYSTEM

The same factors that produced a need for consistent arrangement of building specifications were also operative in the areas of technical data filing and construction cost accounting, and the idea emerged that the requirements of these three activities might be treated uniformly. Action to this effect was initiated by the American Institute of Architects and the Construction Specifications Institute in sponsoring a construction industry meeting to develop a filing system for building product data based on building specifications. This concept was subsequently broadened to include a specification outline and, through the interest of the Associated General Contractors of America, a contractor's cost accounting guide. The results of this activity were published in 1966 as the *Uniform System.*[3] The specification outline embodied in the Uniform system is based on and contains the same 16 divisions established in the CSI Format.

[2]*CSI Manual of Practice,* Construction Specifications Institute (Washington: CSI, 1967) p001/11.

[3]Full title of the publication is *Uniform System for Construction Specifications, Data Filing & Cost Accounting (Title One-Buildings),* issued under the joint copyright of the American Institute of Architects, Associated General Contractors of America, Construction Specifications Institute and the Council of Mechanical Specialty Contracting Industries. Other organizations that have helped develop and/or now endorse the Uniform System are American Society of Landscape Architects, National Society of Professional Engineers, and the Producers' Council. The publication carries individual society or association document numbers as follows: AIA K103; ASLA 3D; AGC 19; CSI 001a; NSPE 1924.

THE SPECIFICATION OUTLINE

The Uniform System embraces the principle that the 16 divisions of the Specification are fixed in name and sequence.

1	General Requirements	9	Finishes
2	Site Work	10	Specialties
3	Concrete	11	Equipment
4	Masonry	12	Furnishings
5	Metals	13	Special construction
6	Carpentry	14	Conveying systems
7	Moisture protection	15	Mechanical
8	Doors, windows, and glass	16	Electrical

Within this framework there can be considerable latitude in assigning section titles and numbers without loss of proximity to related sections, since the fixed divisions assure a relatively constant location for each section within the Specification as a whole. However, as the merits of a uniform system of specification organization become more widely appreciated, the desire for freedom by specification writers to retain the flexibility associated with random designation will no doubt diminish. Furthermore, some standardization of section titles and numbers (or at least sequence) within divisions is necessary for effective utilization of automated typing equipment and electronic information storage and retrieval systems.

Foreseeing this development, the Uniform System suggests that certain "common denominator" sections also be standardized according to titles and sequence whenever a project requires their use. Such sections are called *broadscope*; the less typical are designated as *narrowscope*. Although no numbers have been assigned to sections in the specification outline, a convenient mechanism is available in the cost accounting numbers of the Uniform System. The accompanying table provides parallel listings of some typical sections that might appear in Divisions 2 and 3 of a Specification organized in accordance with the Uniform System recommendations. BROADSCOPE sections appear entirely in capital letters and Narrowscope titles with initial capitals only. The numbers are those established for the cost accounting system.

Division 2: Site Work	Division 3: Concrete
0210 CLEARING OF SITE	0310 CONCRETE FORMWORK
0211 Demolition	
0213 Clearing & Grubbing	0320 CONCRETE REINFORCEMENT

0220	EARTHWORK	0330	CAST-IN-PLACE CONCRETE
0221	Site Grading	0333	Post-Tensioned Concrete
0222	Excavating & Backfilling		
0227	Soil Stabilization	0340	PRECAST CONCRETE
		0341	Precast Concrete Panels
0230	PILING	0342	Precast Structural Concrete
		0343	Precast Prestressed Concrete
0240	SHORING & BRACING		
0241	Sheeting	0350	CEMENTITIOUS DECKS
0242	Underpinning	0351	Poured Gypsum Deck

This description of the specification outline under the Uniform System is synoptic only. For additional detailed information the publication cited in footnote 2 should be consulted as well as CSI Document 001.[4]

GENERAL AND TECHNICAL PROVISIONS

Specifications for building projects contain two classifications of clauses: general provisions and technical provisions. General provisions apply to the work as a whole, whereas technical provisions pertain to technological aspects of the various types of construction, components, and equipment involved. Under the Uniform System the general provisions are placed in Division 1, General Requirements, and the technical provisions comprise Divisions 2 through 16.

General Requirements

As noted in Chapter 4, certain provisions treated broadly in the General Conditions will require modification or additional detailed treatment in the Supplementary Conditions or in Division 1 of the Specifications. Items of a contractual-legal nature should be modified or amplified within the Supplementary Conditions. Items closely related to the contractor's operations at the site should be expanded upon in the general requirements division of the Specifications; under this category, several items identified as broadscope section titles in the Uniform System outline are discussed in the following paragraphs.

SUMMARY OF THE WORK

This is a general summary and description and is normally much less detailed than that included in an invitation to bid or advertisement. It may be considered an introduction to the Specifications and it frequently emphasizes their status as comprising part of the Contract Documents. The general location of the project may be restated and attention called to any special site conditions or restrictions concerning the contractor's operations. It should be stated here whether the

[4]*CSI Format for Construction Specifications,* Construction Specifications Institute, Washington, D.C., January 1967.

work is to be accomplished under a single or under separate contracts; if under separate contracts, the broad divisions covered should be stated. Other items frequently included under this section pertain to work and equipment to be provided by the owner and any portion of the work that is to be postponed until after the designated completion date.

SCHEDULES AND REPORTS

This broadscope title encompasses several diverse items including responsibility for establishing lines, levels, and grades; details of the progress schedule to be prepared by the contractor; schedule of required tests; identification of approved testing agencies; subsurface soil reports; schedule of progress photographs; and similar items relating to the project as a whole.

In subsurface soil reports the accuracy and interpretation of data from borings, test pits, and test piles are a fruitful source of disputes in construction work. Therefore all subsurface data should be excluded from the Specifications and other Contract Documents. Although subsurface information is normally obtained by the owner and architect-engineer, and this is made available for inspection by the contractor, the Specification should state specifically that the owner and architect assume no responsibility for the accuracy of such information or conclusions drawn from it by the contractor and that it does not form a part of the contract.

This section is also the place for making general reference to applicable building standards and standard specifications promulgated by organizations such as the American Society for Testing and Materials, American Concrete Institute, American Institute of Steel Construction, American Society of Heating, Refrigerating and Air Conditioning Engineers. The incorporation of these standards in the technical sections of the Specification by reference is discussed in more detail later in this chapter.

SAMPLES AND SHOP DRAWINGS

Under this heading detailed procedures and schedules for submission of shop drawings and samples are designated, supplementing the provisions contained in the General Conditions. Some specification writers feel that all details in this connection should appear in Division 1 with a reference to this portion of the Specifications included in each of the various technical sections pertaining to work for which shop drawings and samples are required; others concur in the desirability of cross reference, but advocate that information such as number of copies to be submitted in a particular case be stipulated in the technical section concerned with the material or component in question.

TEMPORARY FACILITIES

The two preceding topics pertain to *procedures* relating to the job as a whole; this topic is more directly related to contractor-furnished *items* that apply to job operations as a whole. This encompasses the establishment and maintenance of access roads, barricades, temporary fences, guard rails, scaffolding, staging platforms, hoists, and construction elevators and stairs. Also included in this section are requirements for construction offices, storage for construction materials, temporary toilet facilities, first aid facilities, and drinking water as well as temporary provision of heating, electrical power, and lighting. In addition, security provisions such as watchmen and gate guards are covered here.

PROJECT CLOSEOUT

Under this heading come requirements for record drawings, record specifications, as-built drawings, and setting up procedures for delivering guarantees, bonds, releases, etc. Any special requirements pertaining to handling of the punch list and final inspection that are not included in the pertinent clauses of the General Conditions would also be placed here.

ALLOWANCES

As noted in the General Conditions, the specific amounts of cash allowances should be stipulated in the Specifications. Precise description of the nature of a particular cash allowance should appear in the technical sections pertaining to the component in question; it is recommended, however, that actual dollar amounts be given only in a master list appearing in this section of Division 1. Reference to this portion of the Specification is then included in the technical sections covering the work for which cash allowance is to be made.

ALTERNATES

These fall into two classes: (a) alternates of project scope intended to permit adjustment of bids received so that a contract may be awarded within the budget and (b) alternates of materials and methods that allow selection on the basis of value. Alternates of project scope should be described in detail in this section, whereas alternates of materials and methods should be listed here and specified in complete detail in the various technical sections affected. All alternates should be given an individual number or other reference designation, since each must also be identified on the bid form[5].

Multiplicity of alternates discourages effective competitive bidding and they should be held to a minimum consistent with the particular budgetary situation. The descriptions of alternates in the Specification and their delineation in the Drawings should be so arranged that any one of them, or any combination, may be selected and still have a definitive contract result. Many architects and engineers believe that alternates are more effective when they are set up to be

---

[5]See Chapter 7.

additive (to a base bid) rather than deductive, this practice being strongly recommended wherever conditions permit a choice.

Types of Technical Specifications

There are two basic approaches to writing the technical provisions of building specifications: (a) that based on specifying details of materials and workmanship required and (b) the performance concept. The divergence between the approaches is frequently expressed by the phrase "method *versus* results." In practice most building specifications are a combination of the two, but it is important that the distinctive characteristics of each be understood.

MATERIALS AND WORKMANSHIP SPECIFICATIONS

The term *work procedure specifications* has been proposed as a more appropriate designation for technical specifications of this nature. Whichever term is used, however, the approach consists of specifying in detail the quality, properties, and composition of materials as well as procedures and construction methods to be employed. The basic presumption is that by following the specification provisions the desired result will be achieved. This means that the sufficiency of the design prepared by the architect-engineer must be relied on to produce satisfactory overall performance of the finished component or system. The contractor is committed only to carrying out the construction as called for in the Specification and other Contract Documents; and if his work so conforms and is free from defects, he cannot be held further responsible.

PERFORMANCE SPECIFICATIONS

This type calls for results and prescribed service behavior on completion of a component or system without stipulating the methods by which these are to be attained. Performance specifications have been used in construction work for many years to procure certain items of mechanical and electrical equipment, but their use for general building purposes has only recently begun to expand. Whenever the service performance characteristics of an element of a building can be measured by satisfactory tests, this type of technical specification is appropriate. Under it the contractor agrees to furnish a system or component possessing the specified characteristics and he warrants its future performance in service. Obviously, when operating under performance specifications, the contractor has to be given considerable latitude in the choice of materials and details of assembly with respect to the building element in question. This latitude, however, does not prevent the architect from making the decision between an aluminum, bronze, or stainless steel metal curtain wall, but he would be ill advised to attempt in addition to specify alloys, metal thicknesses, or construction details. To insist on specifying materials, methods, *and* the results to be achieved is unfair and inequitable and may render the architect-engineer liable for rectification costs if the element so specified fails to perform satisfactorily.

The use of performance specifications for mechanical and electrical systems in buildings will undoubtedly increase, particularly as the systems approach to building design in general becomes more widely accepted.[6] The trend toward larger prefabricated components for wall, floor, and ceiling systems also tends to promote the use of technical specifications of this type.

Tests for Specification Compliance

As far as the situation permits, properties and characteristics specified for materials, components, and equipment should be capable of verification by tests. This, of course, is not always possible since some characteristics of building products relating to appearance, texture, etc., do not lend themselves readily to quantitative measurement. In these instances determination of compliance is judged by the architect-engineer on the basis of comparison with samples previously furnished and approved. Standard tests, however, are available for measuring physical, chemical, electrical, thermal, and acoustical properties of materials and components and should be required where applicable.

Many of the standard technical specifications referred to earlier include details of testing procedures which can be made a part of any section of the project technical specifications by reference. Usually, tests of this nature as exemplified by those for measuring strength of concrete, water absorption of brick, absorption coefficient of acoustical materials, etc., fall under the broad classification of quality control tests. In addition, performance tests have been developed to measure the behavior of construction assemblies and components. An example is the one for measuring structural deflection and air and water infiltration of metal curtain walls, promulgated by the National Association of Architectural Metal Manufacturers.

The performance-type test is, of course, most directly applicable to evaluating products procured under performance-type specifications. The development of performance testing may be expected to proceed rapidly with expanding use of performance specifications. The principal difficulty at the present time is that performance tests in general are not necessarily indicative of the service life of a product. This tends to make manufacturers and contractors reluctant to furnish

[6]The most significant development in the United States in this connection is the School Construction Systems Development project (SCSD) carried out in California. In addition to performance specifications covering integrated structural, mechanical, lighting, ceiling, and partition systems, they also extended to criteria controlling functional factors such as flexibility and space requirements. As of early 1967 eleven schools had been built or were under construction. See *SCSD: the Project and the Schools,* Educational Facilities Laboratories (New York, 1967).

warranties guaranteeing durability over extended periods, especially where innovations in building technology are involved.

Judgment must be exercised with respect to the number and severity of tests to be called for. There are many standard products marketed by reliable companies, and under these circumstances certification of compliance with specification requirements may often be accepted in lieu of actual testing for a specific project. This is probably truer in connection with quality control tests than with performance tests, although (to use the metal curtain wall example again) when a standardized assembly produced and erected under the supervision of the manufacturer is involved, extensive performance testing may have been performed before the component was placed on the market. Here certification and warranty might be sufficient. On the other hand, cylinder tests should always be required in connection with poured-in-place concrete, and tests of cores drilled from the hardened concrete may also be desirable in many instances. The question of balance between sufficient and excessive testing and guarantee provisions is very important. Unrealistic requirements can defeat their own ends as well as inflate the bids.

Technological Responsibility

The architect-engineer is presumed knowledgeable in the field of building technology and hence is responsible for the adequacy of the materials, components, and equipment he specifies. Whether specifications are written by a principal of the A & E firm or are prepared by others, the specification writer must select from among products available those best suited to the requirements of the particular project. This involves consideration of the critical properties, performance characteristics, and cost of a material or component as weighed against possible alternatives. The architect-engineer's professional responsibility to exercise reasonable care and skill further requires that a specified product be compatible with adjacent materials and components to assure having the entire assembly function as intended. In making these determinations some guidance may be obtained from the manufacturers' written representations if these are other than sales statements or advertising. Thorough checking of such information is still essential, especially when the equivalence of substitute products proposed by a contractor may be involved[7]. It is also important that nothing

[7]An interesting case history relevent to this point is presented in Bulletin No. 10 in the series issued by the AIA's Committee on Professional Insurance: "An architect, with the assistance of independent heating engineers, designed a radiant heating system for a hospital. Copper tubing was specified, but because of the Korean War became unavailable. The architect accepted a substitute material, relying upon the manufacturer's representation that the new material was suitable and an equivalent substitute for copper tubing. The heating engineers indicated they believed the substitute was suitable and the owner, relying upon the architect, accepted it. The architect issued a change order to provide for this substitute material, but did not secure testing or checking of the new material to see that it was, in fact, suitable and equivalent and did not make any change to adapt the existing design to

appear in specification clauses contrary to a manufacturer's published or supplementary written information relating to proper use and installation procedures.

OWNER'S MATERIALS

When certain materials or products are to be furnished by the owner, their technical adequacy and particularly their compatibility with adjacent materials and components must be given extremely careful consideration. If the materials to be supplied possess characteristics inferior to the quality standards set by the Specification for the project as a whole, this must be made known to the owner. The architect-engineer must avoid placing himself in a situation where he is responsible for the adequacy of the combination of materials specified without having a measure of control over their selection.

Use of Reference Standards

Standard specifications covering specific materials and products are promulgated by technical societies, trade associations, individual manufacturers, and government agencies. They may be incorporated in the project Specification by definitive reference, and when so cited the provisions of the referenced standard have the same status as they would have if reproduced in their entirety. Many of these standards are widely known within the building industry, and consequently products and processes covered by them represent familiar and acceptable usage. In addition, the reference approach reduces the volume of printed matter in the project Specification with consequent saving in preparation time and gain in readability.

In addition to the American Society for Testing and Materials and other organizations mentioned earlier in the chapter, the Portland Cement Association, Acoustical Materials Association, National Electrical Manufacturers Association, and the Structural Clay Products Institute are representative of the many manufacturers and trade associations that issue standard specifications controlling the quality of products produced by their member firms and establishing

the use of the substitute material. During construction, the engineers' foreman suggested to the architect that this substitute material might expand considerably more than copper, but still no provision was made for any expansion joints. A few months after completion, the substitute material had expanded so much that the entire system had to be replaced and the ceiling and floors repaired.

The court said that the architect should not have relied completely on the manufacturer's representation and held that he was negligent in failing to secure independent tests of the suitability of the substitute material to see if it could be used safely in the same manner as copper and without any redesigning."

procedures for their application or installation. In general, such reference specifications are nonrestrictive and their use does not inhibit competition. Although reference specifications prepared by individual manufacturers may be biased, they frequently represent the best current practice for the type of product involved; when using them care must be taken to assure that competitive products are not ruled out by the wording of the particular specification.

Because the U.S. Government initiates such a large volume of construction in the capacity of "owner," the ever-growing number of Federal Specificiations applicable to building design and construction constitutes a major source for reference standards. These cover a wide variety of materials and products[8] and are written in order to facilitate competitive bidding. In addition to the Federal and Military specifications as such, Commercial Standards and Simplified Practice Recommendations are published by the U.S. Department of Commerce which cooperates with various industry groups in developing them.

PRECAUTIONS TO BE OBSERVED

The employment of reference standards as a means of avoiding detailed investigation of the requirements for specifying particular materials, components, or systems is to be greatly discouraged. The technique of using them effectively has been presented in excellent detail in a document issued by the Construction Specifications Institute.[9] It will be sufficient for present purposes to note only that specification writers desiring to make effective use of a reference standard must know the document completely. Many such standards cover a variety of grade, type, or other classification of material and it is necessary to designate specifically which is meant. Otherwise, the contractor may have "hidden choices," thereby vitiating the control the architect-engineer intended to exert.

An additional hazard from the standpoint of construction contract administration results from the statements in many standards that pertain to matters already covered in the General Conditions. This can be a fruitful source of contradiction and ambiguity unless such statements are specifically annulled.

The final point to be made regarding reference standards concerns precise designation of the standard to be referenced. Since many of these documents are under continual study and revision, it is necessary that the specific edition be cited. Some specification writers call for the "latest edition" but the "latest edition" at the time the construction documents were prepared may have been superseded by another before final execution of the contract. Although some individuals feel that citing of reference standards by specific edition is too

---

[8]See *Index of Federal Specifications, Standards, and Handbooks,* U.S. Government Printing Office, Washington, D.C.

[9]*CSI Manual of Practice,* The Use of Reference Standards, Document No. 006.

burdensome a procedure to be practicable, unless this practice is observed, a source of future disputes with respect to enforcement of the Specification becomes built into the Contract Documents.

Closed Specifications

If performance-type technical specifications were available for a wide variety of materials, components, systems, and equipment, and if standard reference specifications existed to cover all other situations where the performance concept is not applicable, there would be little need to specify building products by the use of proprietary trade names. Under such conditions control of quality and performance could be attained by selecting appropriate criteria levels applicable to the various specifications involved. Unfortunately, coverage of this nature is far from complete and the specifier frequently must resort to naming proprietary products with whose quality and performance he is familiar. Sometimes he may know of several products that will serve the required purpose, but often there will be only one brand which, in his experience, will meet the specific conditions of the particular project. This leads to a closed specification which may be defined as one that limits acceptable products to one or a few brand-identified types or models and prohibits substitutions.

Although achieving project technological control, the noncompetitive nature of this procedure (particularly where only one product is named) seldom results in the most favorable price to the owner. The situation is mitigated somewhat when three or more comparable proprietary products can be identified with any one of them acceptable. This method of specifying is common in private construction work and some public agencies permit its use under these conditions. However, most public construction is required by law to be bid under open specifications as described in the following discussion.

Open Specifications

In general, technical specifications of the performance type as well as those of the quality or product description type represented by the reference standards lend themselves readily to open specification requirements. The principal characteristic of an open specification is that procedures are provided to accept the products of various manufacturers whether or not they are mentioned by name, providing such products can meet the stipulated requirements, that is, substitutions are permitted. The principal concern of the specifier regarding open specifications relates to methods for handling substitutions so that adequate control over quality and performance can be maintained by the architect-engineer and/or the owner. In general, open specifications are characterized by the listing of one or more proprietary names for a product followed by the phrase "or equal."

THE "OR EQUAL" PROBLEM

These words, which have probably generated more debate than any others in the specification writer's vocabulary, immediately raise the question of how equality is to be established and who will be the judge. When the phrase is used, it is essential that the Conditions of the Contract or the General Requirements provisions of the Specification explain this point explicitly. This is usually accomplished by a statement similar to the following:

> Wherever used in these Specifications, the phrase "or equal" means that materials, components, and equipment proposed for use in lieu of those named will be considered acceptable if they will, in the opinion of the architect-engineer, perform adequately the functions imposed by the general design, and if they meet the minimum standards in all respects of those of the named item.

Suggestions have been made that phrases such as "or approved equal," "or approved alternate," "or equivalent alternate," etc., would be more appropriately related to the substitution procedure than "or equal." Semantic exercises of this nature, however, do not really clarify the situation since some similar definition as that given above is required. State laws governing public works construction frequently are explicit in this regard. California prohibits the specifying by brand or trade name ". . . unless the specification lists at least two brands or trade names of comparable quality or utility and is followed by the words 'or equal' so that bidders may furnish any equal material, product, thing, or service."[10] There is, of course, no similar rigidity in private construction except as it may be imposed by the owner.

Although the "or equal" approach to specification policy has received much justifiable criticism, it is still more commonly used in practice than any other. It will be noted that as long as any substitutions are to be permitted, a mechanism must be provided for handling them. In private construction restrictions concerning substitutions can be invoked for those portions of the Specification where it appears in the owner's interest to do so. It is questionable, however, whether a closed Specification for the bulk of the project is of ultimate advantage to the owner.

HANDLING SUBSTITUTIONS

Control of substitutions is without doubt one of the most difficult problems faced by the architect-engineer. There is the responsibility for making the technological decision as to whether a proposed substitute product will, in fact, provide the same in-service performance as those named in the Specification and with whose qualities and service characteristics the architect-engineer is familiar. Procedures for handling substitutions must place the responsibility on the

[10]Sect. 4380 of Article 4, Government Code of the State of California.

contractor to prove to the satisfaction of the architect-engineer that a substitution offered meets the quality, fitness, and capacity of the item originally called for. The contractor should be required to furnish complete performance data, descriptions, test reports, and similar supporting information on which the decision to accept or reject a proposed substitution can be made. Since architects and engineers can be held liable for subsequently discovered inadequacy of materials, components, and equipment specified or approved as substitutions, such decisions must be made with care.

The processing of substitution requests can become a most time-consuming operation for the A & E office since, if carried to extremes, it virtually amounts to redesign of many portions of a project. As noted earlier in the discussion of technological responsibility, it is necessary that products specified be compatible with adjacent materials and components for the assembly as a whole to function satisfactorily. Reliable information in this connection is normally lacking, and this is one of the principal factors creating reluctance to approve substitutions in the absence of first-hand experience. In addition, many design offices feel that most substitutions result in compromise of intended quality.

In private work it is possible to arrange that proposals for substitutions be made and accepted or rejected before the construction contract is awarded. In much public construction, however, this is not feasible; for example, the California law cited requires that "Specifications shall provide a period of time of at least 35 days after award of the contract for submission of data substantiating a request for a substitution of 'an equal' item." This would appear to be a maximum period that is consistent with efficient administration of the project. Wherever an option exists the period should be made substantially less, so that in the event a substitution request is disallowed, delay in procurement of the item concerned will not cause delay in the progress of construction.

Combination Type Specifications

Several schemes have been devised for writing specifications for private work that attempt to combine the merits of the open and closed concepts but minimizing their respective disadvantages. In all of these attempts the primary concern is to discourage proliferation of substitution proposals, thereby facilitating technological control by the architect-engineer. Unfortunately, a generally accepted terminology has not yet been developed for the several schemes in current use. The Construction Specifications Institute identifies seven of these and has assigned tentative names to each.[11] The following discussion focuses on

[11]*CSI Manual of Practice,* Proprietary Specifications in Private Work, Document No. 008. This document gives descriptions of the several types and also presents much additional comment on their relative merits.

two procedures intended to encompass some of the advantages of the competition provided by open specificaitons while retaining a large measure of control characteristic of the closed type. The first pertains to situations where admissibility of substitutions will be resolved before receipt of bids and the second where the matter is to be decided after bids are received but before the contract is awarded.

PRODUCT APPROVAL SPECIFICATIONS

This scheme is sometimes known as the "Or Equal with Prior Approval" technique. Under it the technical specification lists one or more proprietary brands for each product called for with a provision that any proposal for substitution by a bidder must be made a specified number of days before bid opening. All approvals of substitute products are then issued to all bidders of record by means of an Addendum. After execution of the construction contract, no further proposals for substitutions are in order and the contractor must furnish one of the products listed in the Specification or covered by the Addendum. In order to prevent a deluge of requests for product approval by manufacturers' representatives, it is prudent to limit submissions to only those prime bidders who wish to make the substitution. Naturally, all substitution proposals should be accompanied by the same supporting performance data, test reports, etc., as would be required under an open specification.

It should be observed that many architects and engineers doubt whether this method of handling substitutions actually accomplishes the ends intended. To be workable the bidding period must be sufficiently long to provide time for adequate review and decision, and this kind of time is not always (or even frequently) available. Some feel that if consideration of substitutions is deferred until the successful contractor is known, the number of requests will be less than if submitted by several bidders during the bidding period; furthermore, more time will then be available for arriving at a sound decision concerning the merits of proposed substitutions. Concurrence or disagreement with this appraisal of the scheme will vary among architect-engineers by virtue of individual experiences. If sufficient options are provided in the project Specification, proposals for substitutions subsequent to award of contract may be minimal, especially if any savings due to lower prices must be credited to the owner.

SUBSTITUTE BID SPECIFICATIONS

In this scheme bids are received on the basis of the one or more propriety products listed for any particular item as in a closed specification, and the contract award is made on this "base bid" basis. At the time of submitting bids, however, a bidder may file with his bid a proposal for a substitute product and show the amount to be added or deducted from his base bid if the proposed substitution is accepted. An advantage of this procedure is that the contractor may submit an alternate material or product he prefers to use and the architect-engineer can evaluate its merits against the additional cost or savings involved. These decisions should be made before execution of the construction contract so that the substitutions can be made a part of the contract.

It is essential that award of the contract be made on the base bid without consideration of the substitution proposals. Otherwise there is the risk of becoming hopelessly involved in the question of which bidder was actually low and of generating suspicion that the award was manipulated to favor a previously chosen contractor. One undesirable factor of this procedure, however, is that little incentive exists for a bidder to prepare substitute proposals, and many architects and engineers have found that few bidders make such proposals. Although this is satisfactory from the standpoint of discouraging substitutions, it is less satisfactory in preserving some of the competitive aspects of open specifications.

Nevertheless, where there is a choice in private work, an effort to control unlimited substitution proposals is very much worthwhile. Many owners, recognizing that they have retained architects and engineers partly for their judgment in matters relating to technological sufficiency of the building components, systems, and equipment they recommend, may be persuaded that provision in the Specification for unlimited submission and evaluation of substitutions is not in their own overall interest.

Language of the Specification

As noted earlier, although the Drawings and Specifications are complementary with no intended order of precedence between them, there actually is a *de facto* precedence in favor of the Specification because it is a written document and consequently more generally understood than are drawings. Clarity and precision in specification writing are therefore of significant importance. Ambiguity can be costly to the owner, since in disputes over intended meaning of contract provisions the courts usually rule against the party that prepared the contract.

Specifications should be written in short, concise sentences with simple, common words used in their commonly understood meanings. Some technical terms are, of course, necessary, but these should be kept within the range of knowledge of the building trades workers involved. Complicated sentence structure that depends on punctuation to be interpreted correctly should be avoided; misplaced or omitted punctuation can result in a meaning quite different from that intended. With the growing relative importance of the Specification as one of the Contract Documents, skillful composition of specification clauses becomes of increasing significance.[12]

[12]For more detailed discussion of this aspect of specification writing see R.W. Abbett, *Engineering Contracts and Specifications* (New York: Wiley, 1963), Chapter 18; and *CSI Manual of Practice,* The Language of Specifications, Document No. 005.

It is especially important that terms used on drawings for the identification of items be identical with those employed in specification clauses. The "foyer" shown on the ground floor plan should not become a "vestibule" in the ceramic tile specification; nor should the "parking area" in the asphalt paving specification be identified as "parking apron" on the site plan. There are endless opportunities for discrepancies of this nature and they constitute a troublesome problem of coordination. Unless resolved, however, uncertainties are created in the minds of users of the documents, which may become a source of future disputes. Practicable office procedures are available for handling this problem, but the importance of doing so warrants more general recognition and emphasis.

Use of Scope Statements

One of the most controversial points relating to composition of the Specification is whether or not a paragraph summarizing the "scope" should be given at the beginning of each technical section. The inclusion of "scopes" is a custom of long standing which had considerable merit when lengthy "trade sections" were the rule in specification writing. They served as a combined introduction and summary of work dovered in the section. However, with the advent of the CSI Format and the Uniform System, both advocating shorter and more numerous technical sections, the value of scope paragraphs and the advisability of using them have been questioned.

Since every specification section is actually an article of the construction contract, it should be read by all concerned with it; consequently, there is little reason for a scope paragraph that tells what is or is not included in the section. Three principal topics traditionally (and often still) treated in scope paragraphs deal with the relation of the particular section to the Conditions of the Contract, the work included in the section, and work not included. Since the use and phrasing of these provisions in the Specification may have considerable impact on effective administration of the construction contract, they are considered in more detail below.

REFERENCE TO CONDITIONS OF THE CONTRACT

It is the opinion of some specification writers that a statement such as the following should preface each section in Divisions 2 through 16:

> The requirements of the General Conditions, the Supplementary Conditions, and Division 1 of these Specifications apply to all work under this Section.

The feeling that such a general reference is necessary doubtless survives from earlier practice when there was an attempt to establish the then longer specification section as co-extensive with a subcontract. Although the subject matter of any of the shorter single sections characteristic of current practice should not relate to the work of more than one subcontract, the combination of sections (that is, "units of work") into subcontracts is the responsibility of the prime contractor. In fact, some architects and engineers emphasize this point by stating in the general provisions of Division 1 that neither the arrangement of subject matter nor titles and headings of sections and paragraphs are to be interpreted as correct or as complete segregation of the various kinds of materials and labor required. Under these conditions the value of a reference statement in individual sections to alert potential subcontractors to the existence of these other elements of the construction contract seems questionable.

Furthermore, it may be inadvisable since the prime contractor, by terms of the construction contract, must bind all subcontractors under conditions similar to those binding him to the owner. This means he is legally bound to inform subcontractors about provisions of the Conditions of the Contract. The fact that under current operating practices in the building industry the prime contractor may sometimes fail to do so is not a sound reason for the architect-engineer to assume this responsibility. It is now generally recognized that all portions of a construction contract, as represented by the Contract Documents, are applicable to all parties concerned. Consequently, an attempt to indicate in the Specifications certain portions that are "more applicable" than others may weaken the intended balance among the several documents and work to the disadvantage of the owner and architect-engineer in event of future disputes. Rather than attempting to clarify matters of this nature for a subcontractor (a third party since he is not a party to the contract between owner and prime contractor), it is more prudent to emphasize the obligation of the prime contractor by incorporating a clause similar to the one below in Division 1 of the Specification:

> Since the Conditions of the Contract and the General Requirements (Division 1) of these Specifications apply to the work described under each Section hereof, the Contractor shall instruct each of his Subcontractors to become fully familiar with them.

Incorporating copies of the General Conditions and Supplementary Conditions in the Project Manual or Construction Documents Book, as discussed earlier, facilitates this action.

WORK INCLUDED

Whatever merit statements purporting to describe or enumerate items of work included under a specification section may have had in the past, there is no valid reason for using one under current conditions of practice. It is the drawings that should be relied on to show the extent of all of the work, since it is impossible that such listings can ever be complete. If they are given at all they may imply that all work is listed, thereby creating a source of future disputes. The hedging phrase ". . . work includes but is not limited to the following listing . . ." is by no means a reliable protection against incomplete lists; if it were, there would be little reason for making such lists.

Another hazard in exhaustively "complete" scope statements is the duplication of information shown on the Drawings. If changes are made in the Drawings and are not indicated in the scope paragraphs, discrepancies are created between the Drawings and the Specifications. These errors are easily generated whenever statements in the specification sections "tell what is on the drawings," and scope clauses are in effect largely of this nature. A safer policy is to omit them, thus forcing the bidder to read the complete section.

WORK NOT INCLUDED

This is a puzzling title that still appears in some specification sections. Its intended purpose is clarification, that is, to indicate that some work that might possibly be expected to appear in the section is excluded and covered elsewhere. Both the need for information of this nature and the appropriateness of supplying it under current conditions of practice are questionable. It is relevant only if the specification section is treated as a mechanism for controlling the scope (work included and work excluded) of subcontracts; this premise is not valid in the complex labor relationships and the complex technology that characterize the building industry today. As noted, the assignment of work under specification sections to subcontractors is a function and responsibility of the prime contractor.

There is also the risk of ambiguity in the construction contract when items shown on the Drawings are listed under "Work Not Included" articles and their treatment elsewhere in the Specifications is inadvertently omitted. Under such conditions a contractor may be able to sustain a claim that he was justified in omitting the listed items from his bid even though the General Conditions provide that the Drawings and Specifications are complementary and what is required by one is as binding as if required by both. Whatever value information under this title may have for purposes of clarification and cross-reference is outweighed by its potential for causing trouble under current conditions of practice. Furthermore, the trend toward use of shorter sections advocated by the Uniform System will eventually eliminate the feeling of any need for scope statements.

Chapter 6

Bonds and Construction Insurance

The general nature of surety bonds and construction insurance, as devices to assure completion of the project and protect the interests of all parties concerned during the construction process, was pointed out in Chapter 4 under the discussion of articles of the General Conditions. This chapter deals with these most significant factors of construction contracting practice in more detail.

SURETY BONDS

The signal characteristic of surety bonds is that they provide protection against loss from failure of a third party to perform stipulated obligations he has undertaken. As related to building construction, the bond is an agreement by the bonding company (surety) supplying it to indemnify the owner (obligee) for nonperformance by the contractor (principal). The extent of indemnity for which the surety is liable is limited by the amount stated in the bond and known as the face value or penal sum. In effect, a surety bond backs up the financial responsibility of the contractor for the benefit of the owner to the extent of the penal sum.

Performance Bond

This bond guarantees performance in accordance with the terms of the construction contract. Customarily it is a simple document and does not designate specific responsibilities of either the contractor or the surety. Obligations under the bond are identical with those of the construction contract and hence any inadequacies in the contract will constitute similar inadequacies in the bond. If

189

the surety has to take over upon default of the contractor, it is obliged to perform the work in accordance with the Contract Documents to the amount of the penal sum of the bond. Since the surety has no liability for costs exceeding this amount, it is recommended that performance bonds be written to cover 100% of the contract price. In event of default the surety may proceed according to its own judgment about the best way to complete the contract. It may engage another contractor to do so, may employ the forces of the defaulted contractor, or may accomplish the uncompleted portion of the work with its own forces.

Where the construction contract requires maintenance obligations over a warranty period of one year following completion, the obligation of the bond also runs to the end of the same period. Some states have statutory provisions relating to surety bonds and these must, of course, be followed where applicable. The performance bond form recommended by the American Institute of Architects is reproduced at the end of this chapter and illustrates the arrangement and provisions of bonds used in private work where a statutory form is not prescribed.

Payment Bond

Known more definitively as a Labor and Material Payment Bond, this bond guarantees that the contractor's bills for labor and materials incurred under the contract will be paid. The existence of such guarantee protects the owner from liens and other claims made after completion of the project and after final payment has been made to the contractor. The relationship among the parties concerned is somewhat more complex than with performance bonds since payment bonds serve the function of protecting subcontractors, material vendors, and workmen although the owner is still the obligee.

Early experience in the use of labor and material payment bonds demonstrated the necessity for excluding claimants unreasonably remote from the prime contractor and for qualification of the terms labor and material. It will be noted from perusal of the AIA payment bond form reproduced at the end of the chapter that the first point is covered by defining a claimant as one having a direct contract ". . . with the Principle or with a subcontractor of the Principle for labor, material, or both . . .", and the second by construing labor and material to include ". . . water, gas, power, light, heat, oil, gasoline, telephone service or rental of equipment directly applicable to the Contract."

It is essential that the procedures stipulated in the bond pertaining to notification, time limitation, and court jurisdiction be followed explicitly when action by a claimant is instituted.

The Dual-Bond System

Performance bonds were the first of the construction contract bonds to be developed. Subsequently, provisions were incorporated to underwrite payment for labor and material; from these the combination performance and payment bond evolved. The single combination bond, however, proved to have serious drawbacks in practice since there was a built-in conflict of interests between the owner and those who furnished labor and material. Action on claims of the latter frequently was delayed pending settlement of the owner's claims which had priority. These difficulties led to developing the dual-bond system under which the interests of the owner and those of subcontractors, material vendors, and workmen are covered separately. This two-bond pattern is exemplified by the forms at the end of the chapter which constitute AIA Document A311 and are widely used in private construction work.

The dual-bond system has been mandatory on federal construction projects since passage of the Miller Act by Congress in 1935. Before passage, subcontractors and others who furnished labor or material on federal projects were in a more vulnerable position than when engaged on private work since generally liens cannot be filed against public property.[1] The two-bond pattern is implemented through use of *Standard Form 25, Performance Bond* and *Standard Form 25A, Payment Bond* prescribed by the General Services Administration. The Miller Act permits the claimant to sue on the payment bond under prescribed procedures in the appropriate U.S. District Court. Here, as in private construction work, protection is provided those who deal directly with the prime contractor and those who deal directly with a subcontractor who has a direct contractual relationship with the prime contractor.[2]

Surety Bonds and the Contract

Since construction contracts always provide that the owner may order changes in the work without invalidating the contract, a question arises as to whether such changes may invalidate a performance bond. By terms of the bond the surety guarantees performance in accordance with the obligations stated in the construction contract at the time both are executed. However, since the surety is not a party to the construction contract, changes made by owner and contractor appear to bind the surety to contractual agreements made without its consent.

---

[1]However, some states allow filing a lien against money earned by but not yet paid to a contractor on public construction.

[2]For a discussion of the rights of sub-subcontractors in public construction with respect to payment bonds see H.A. Cohen, *Public Construction Contracts and the Law* (New York: McGraw-Hill, 1961), pp. 208-213.

Such a situation might discharge the bond, although there is opinion that as long as the original contract contains specific provision for changes, the surety is obligated to accept such modifications.

In order to eliminate uncertainty in this connection many bond forms incorporate a statement that the surety waives notice of any alteration or extension of time made by the owner. Provisions of this nature are mandatory in federal government construction projects. Even with this statement included, substantial changes in the scope of a project should not be undertaken without notifying the surety. At some point the scope of a change may appear sufficiently great to constitute in effect the making of a different contract; nevertheless, there will be some limit to expansion of the surety's obligations under the change order mechanism that the courts would uphold.

It must be borne in mind that the surety has relied on the terms of the contract between owner and contractor in appraising the risk when furnishing the bond. Consequently, it is fully as important that the owner strictly observe its terms as does the contractor. Changes in the handling of retained percentages or payment schedules should never be undertaken without consent of surety, and the concurrence of surety must be obtained before making final payment to the contractor.

Choice of Surety

Although the owner should reserve the right to approve the contractor's bonding company, it is not advisable to require that bonds be furnished by a specific surety. The architect-engineer's concern on the owner's behalf is knowing that the organization from which bonds are purchased has adequate financial resources, a sound reputation for experience and competence in the construction suretyship field, and a record of prompt and equitable action in carrying out its obligations. Many contractors have established long-time working relationships with bonding companies and such sureties possess intimate knowledge of the contractor's financial position, current commitments, equipment resources, and personnel. This information is reviewed and weighed before underwriting additional bonds in order to avert extension of a contractor's operations beyond his capabilities. If all sureties were guided by this precept, there might be little to differentiate among them. Unfortunately, in the competition to sell bonds, some agents may stop short of making as searching an investigation of the facts as prudent underwriting implies; hence the importance of the surety's reputation.

As with insurance companies, surety corporations are subject to public regulation. They must secure licenses to operate in states other than the one in which they were incorporated and must file premium rates and other data with designated public agencies. To furnish bonds for federal construction contracts a surety must also meet financial qualifications prescribed by the U.S. Treasury

Department. A list of companies meeting these qualifications is maintained by the Department and gives the underwriting limit of each as well as the states in which each company is licensed to operate.

Surety bonds play a vital role in building industry operations and an owner should be urged to consult his legal counsel before deciding to enter into a construction contract without one. Scrutiny of a contractor's position by surety before furnishing bonds is a safeguard not to be waived lightly.[3] A further significant aspect of the relationship between contractor and surety with respect to prequalification of bidders is pointed out in Chapter 7.

Bid Bonds

The subject of bid security as a factor in establishing bidding and award procedures is considered in Chapter 7 with discussion here limited to the character and function of surety bonds used for bid security.

A bid bond is a guarantee on the part of the surety that the bidder (principal), if awarded the contract in accordance with conditions described in the bidding instructions, will, in fact, execute the contract with the owner (obligee) and furnish the performance and labor and material payment bonds required. If the successful bidder does not carry out these obligations, the surety is liable for an amount (not in excess of the penal sum) equal to the difference between the successful bid and that of the next higher responsible bidder with whom the owner can contract for the work.

It should be noted that a bid bond assures the owner only of securing a construction contract for the amount stated in the successful bidder's proposal. It does not guarantee that he will get a building for that price. Assurance of such nature is the function of the performance and payment bonds. AIA Document A310 reproduced at the end of the chapter is representative of bid bond forms used in private construction. Federal government contracts use a document designated *Standard Form 24* prescribed by the General Services Administration.

Miscellaneous Bonds and Warranties

In addition to the three types of surety bonds discussed there are several others whose use is less general. Among these are bonds to discharge liens that have been filed against an owner's property by persons who have not received payment for labor or materials supplied; bonds covering release of the retained percentage where statutes prohibit such release before full completion of the contract unless bond is furnished by the contractor; and license or permit bonds

---

[3]For an excellent discussion of the relationship between surety and contractor, see R.H. Clough, *Construction Contracting* (New York: Wiley, 1960), pp. 143-146.

called for by state law or municipal ordinance where building contractors are required to be licensed.

Warranties are closely related to surety bonds, but serve a distinct function. By means of a warranty a contractor, subcontractor, or manufacturer certifies that the material, product, or equipment in question will perform in accordance with the specified requirements. Although a warranty does not necessarily have to be in writing, it is common specification practice to require that written warranties be delivered to the owner at the time of final payment. A distinguishing characteristic of warranties is that they bind the producer or supplier directly without interposition of a surety. A common example of written warranty is the widely used guaranty for bituminous roofing sometimes referred to as a "roof bond." Most written warranties carefully circumscribe the conditions under which the supplier will repair or replace his product and therefore should be scrutinized to determine the degree of protection actually provided.

CONSTRUCTION INSURANCE

The purpose of construction insurance is to protect the financial status of owner and contractor from risks involved during the course of project construction. The architect-engineer has more than an indirect interest in the matter, since he normally participates in administration of the construction contract although he is not a party to it. The function of professional liability insurance as protection in this regard was discussed in Chapter 2; its relationship to contractual liability insurance, carried by the contractor under his indemnification obligations stipulated in the General Conditions, will be apparent. The provisions for construction insurance contained in standardized documents must necessarily be broad in nature, since risks will vary regionally as well as with the type of project. Consequently, the hazards to be covered and the amounts of insurance required should be stipulated in the Supplementary Conditions.

The owner should make the decision concerning construction insurance coverage with the advice of legal and insurance counsel. The architect-engineer in assisting the owner to develop the construction contract may, of course, contribute recommendations from his own experience, but he is not in a position to give professional advice on insurance matters. The three broad areas to which consideration should be directed are contractor's liability insurance, project and property insurance, and owner's liability insurance.

Contractor's Liability Insurance

Under this coverage the contractor purchases and maintains protection from claims that may arise or result from his operations under the construction contract, including the operations of his employees, subcontractors, and others for whom he may be responsible. The insurance encompasses these five areas of the contractor's liability:

Workmen's compensation
Employer's liability
Public liability
Property damage liability
Contractual liability

This listing does not imply that five separate insurance policies are necessarily required, since combinations of some of them are available in comprehensive or package-type policies.

Workmen's Compensation Insurance

Detailed requirements for this type of insurance are prescribed in the workmen's compensation laws and other employee benefit acts of the several states. Benefits to an injured worker may include hospitalization, medical treatment and maintenance payments during convalescence as well as allowances for disablement and, in the event of death, benefits for the worker's dependents. Under the workmen's compensation laws of most states the injured employee is barred from suing his employer for damages in excess of the amount awarded. Thus a concrete worker injured through the collapse of formwork could not additionally sue the contractor on the grounds of negligence. This restriction has generated the filing of law suits against architects and engineers by individuals unhappy with the amount of their award under workmen's compensation on the grounds of negligence in "supervising" the operations of the contractor.[4] The fact that some suits of this nature have been successful has led to incorporation of indemnification or "hold harmless" provisions in the General Conditions as discussed under contractor's contractual liability.

Employer's Liability Insurance

This insurance protects the contractor against claims of employees who are not covered by workmen's compensation laws. It is usually written in conjunction with workmen's compensation insurance and is broad in scope. Coverage extends to bodily injury or death, occupational sickness and disease, and personal injury claims such as shock, mental anguish, restraint, and invasion of privacy. In addition, it performs a contingent function for the contractor when an employee under workmen's compensation coverage seeks further relief beyond an award he deems insufficient. Since he is barred from suing his employer, he may seek

[4]For an illuminating discussion of this situation see the four-part article by Bernard Tomson and Norman Coplan in their column "It's the Law," *Progressive Architecture,* October, November, December 1966 and January 1967.

other targets such as the architect, a subcontractor, or even the owner. When this occurs, the defendant usually takes immediate steps to make the contractor a party to the action. Employer's liability insurance provides for paying the legal costs incurred in his defense and any judgment that may be declared against him.

Public Liability Insurance

Construction operations present hazards to persons other than the contractor's employees. During the course of the work many individuals associated with the project in different capacities will be on the site, and the contractor's responsibility for personal injury extends to them as well as any person in the vicinity who may not be connected with the work in any way. The purpose of contractor's public liability insurance is therefore to protect him against claims for damages resulting from personal injury, sickness, disease, or death of any person other than his employees. In this connection he also may be held jointly liable for damages arising from the acts of his subcontractors even though they carry their own public liability insurance. Protection against this contingent liability is provided by policies issued under the title of Contractor's Protective Liability Insurance or Contractor's Contingent Public Liability Insurance. Such coverage also protects the contractor against liability which may be imposed by law for damages resulting from omissions or supervisory acts by him in connection with the performance of work by a subcontractor.

Property Damage Liability Insurance

Policies of this nature relate to damage to property of others resulting from the contractor's operations. Property owned, leased, rented, or under the custody or control of the contractor or his employees is excluded from this coverage. The purpose of this type of insurance is to cover claims for damages resulting from injury to or destruction of tangible property, including loss of use because of such injury or destruction. Since many standard forms of policies for property damage liability insurance exclude certain types of claims unless special endorsement is incorporated adding such coverage, careful study must be given to establishing the scope of protection for individual projects. Types of claims generally excluded are those arising from demolition work, blasting, and explosion; collapse caused by excavation, shoring, and underpinning; and damage to underground utilities. When any of these risks are known to be present, the pertinent exclusions should be eliminated by endorsement. Correspondingly, specific risks covered in a standard policy, but which are known not to be present in an individual project, may be excluded by endorsement. Obviously, such modifications of policy coverage will be reflected in the premium cost.

Contractual Liability Insurance.

As noted in the article on responsibilities and rights of the contractor in Chapter 4, the growing number of lawsuits against owners, architects, and engineers arising out of jobsite accidents has led to incorporation of indemnification and hold harmless clauses in construction contracts. Assumption by the contractor of liability as described is outside the scope of liability imposed on him by law and consequently outside the coverage of the usual forms of contractor's liability insurance. Whenever indemnification obligations such as those embodied in the standard General Conditions of AIA Document A201 are assumed by the contractor, a special policy or endorsement covering the contractual liability is required. The contractor should make certain that the scope of such hold harmless provisions is, in fact, matched by the coverage of the insurance. However, policies are generally available that adequately meet protection requirements for the contractual liability imposed by AIA Document A201.

Completed Operations Liability Insurance

The purpose of this type of insurance is to extend liability coverage for a specified period beyond the date of completion and acceptance of the project by the owner. Unless specific endorsement is made otherwise, the contractor's liability protection expires at the end of the job, although he continues liable for damages for accidents arising from his operations but occurring after their completion. Whether or not such completed operations coverage should be provided and if so, for what length of time, depends on the type and occupancy characteristics of the individual project. Since cost of this and that of all contractor's liability insurance will be reflected in the bids, this decision should be made prior to soliciting proposals.

Project and Property Insurance

The purpose of this insurance, purchased and maintained by the owner, is to furnish protection to the project against the hazards of fire, extended coverage, vandalism, and malicious mischief. The amount should be sufficient to cover the full insurable value of the entire work at the site and should protect the interests of the contractor, subcontractors, and sub-subcontractors as well as that of the owner during the process of construction. Property insurance subsequent to completion and acceptance of the project is, of course, a continuing responsibility of the owner. A special characteristic of this type of insurance is that it protects the physical aspects of the construction operation from the hazards identified; the cost of replacing or repairing elements of the construction so damaged is assured whether or not title to that portion of the work has passed to the owner. Provisions may vary with respect to extent of coverage for temporary sheds, scaffolding, towers, and other equipment, but the standard policy forms now available in most states contain legally approved standard definitions of items constituting or excluded from the scope.

The existence of project loss and damage insurance is so important to the contractor and those associated with him under the construction contract that the standard AIA General Conditions provide the following:

"The Owner shall file a copy of all policies with the Contractor before an exposure to loss may occur. If the Owner does not intend to purchase such insurance, he shall inform the Contractor in writing prior to commencement of the Work. The Contractor may then effect insurance which will protect the interests of himself, his Subcontractors and the Sub-subcontractors in the Work, and by appropriate Change Order the cost thereof shall be charged to the Owner."

If, on the other hand, the owner desires to have this insurance coverage effected by the contractor, such provision should be stipulated in the Supplementary Conditions by ammending the pertinent clauses of the General Conditions.

TYPES OF POLICY

The two types of policy available to cover loss or damage to the project are the Fire Insurance with Extended Coverage and the All Physical Loss. The first type was developed originally to cover fire losses only and was then expanded to include certain other physical risks. The "extended coverage" of current standard policies of this type generally provides protection against wind, hail, explosion, riot, civil commotion, smoke, vehicles, and aircraft damage. This may be further extended to cover hazards such as vandalism, malicious mischief, and water leakage. The All Physical Loss form sets up substantially the same coverage but on a broader basis; thus specific exclusions can be made where circumstances surrounding a particular project so warrant.

It should be noted that special endorsements are required on policies of either type where insurance against earthquake damage, nuclear contamination, floods, seismic sea waves, and similar disaster-producing phenomena is desired. If appraisal in the light of geographic and physical characteristics of the place of building so indicates, provision for insuring such special risks should be included in the Supplementary Conditions.

BUILDER'S RISK COMPLETED VALUE FORM

Coverage for the several risks discussed may be provided by a single comprehensive policy commonly known as Builder's Risk Insurance. One of the factors peculiar to insurance of this nature is that the value of the work protected varies from nearly zero at the start of construction operations to a maximum on the date of completion. The intent of the insurance, however, is to cover the full insurable value at any one time. This is accomplished readily by the Builder's Risk Completed Value Form which is based on the final completed value of the project, but the coverage at any specific time is limited to the actual insurable value of the project at that time. It should be noted that the "insurable value" is less than actual value since the cost of foundations, excavation, and underground utilities is not normally included.

ADJUSTMENT OF LOSS

Since insurance against damage to the project covers the interests of owner, contractor, subcontractors, and sub-subcontractors, it is necessary that procedures be established for handling claims of the several parties in event loss occurs. The AIA General Conditions provide that any loss covered by the policy will be adjusted with the owner and paid to him as trustee for the others in accordance with their interests. To facilitate this procedure, the owner and contractor waive rights against each other for damages caused by any of the insured hazards, and the contractor agrees to require similar waivers by subcontractors and sub-subcontractors. The owner as trustee has the power to negotiate settlement with the insurance company unless one of the parties concerned objects to his doing so, in which case the matter goes to arbitration. Settlement and distribution of insurance payments are then accomplished in accordance with the directions of the arbitrators.

SUBROGATION

The principle of subrogation may be defined as the substitution of the insurance company for the owner with regard to owner's rights to proceed against third parties. On payment of a loss under the policy, the owner's right of recovery from any persons legally liable for the damage passes to the insurance company. Thus, if fire damage to the project occurred through negligence on the part of the contractor, the insurance company could sue the contractor although the contractor thought he was protected from loss by the owner's fire insurance coverage. The implementation of the right of subrogation could therefore entirely defeat the intended purpose of project-loss insurance. The provisions relating to construction insurance contained in the AIA General Conditions are intended to minimize the possibility of such action. Alternative procedures are to name the contractor as one of the insured under the policy or to seek waiver of subrogation rights by the insurance company. Since the legal validity of such waivers may be uncertain in some jurisdictions, legal counsel should be sought to assure effectiveness of arrangements that are intended to regulate exercise of the right of subrogation.

LOSS OF USE INSURANCE

In many cases the dollar amount of loss to the owner due to delayed occupancy may equal or exceed the cost of the physical damage sustained. Such loss of use constitutes an insurable hazard, and it should be considered whether this coverage is warranted on any particular project. It is, of course, purchased and maintained by the owner.

Owner's Liability Insurance

During the course of construction this type of insurance is in the nature of contingent liability coverage for the owner. It is also known as Owner's

Protective Liability Insurance and provides protection against claims for personal injury and death, or damage to property other than the project, resulting from accidents during construction operations. Although the contractor and his sub-contractors are directly responsible and liable for their operations (and presumably covered by their own liability insurance), the owner is often named in suits brought against them for alleged negligent acts. This insurance also protects the owner against legal liability that may be imposed because of his own supervisory acts in connection with work performed under the construction contract.

The owner's contingent liability insurance duplicates to some extent the protection provided by the contractor's indemnification obligations and covered by his contractual liability insurance. Duplication of premiums for the same protection can be eliminated where both policies are written by the same company or where, as under cost-plus-fee construction contracts, it is customary to furnish this contingent protection for the owner by endorsement to the contractor's insurance. Irrespective of any public liability insurance the owner may have in force on his property before start of the project, he should have contingent liability coverage before work is begun at the site. Maintaining public liability insurance after completion and acceptance is, of course, a continuing responsibility of the owner.

Effecting Insurance Coverage

It was noted earlier that the actual decision concerning type and amount of construction insurance coverage should be made by the owner with advice of insurance and legal counsel. The question is usually raised first, however, by the architect-engineer during the construction document phase of A & E services. To initiate action some firms write the owner calling attention to the fact that the Conditions of the Construction Contract will require the contractor to carry certain types of insurance and requesting his written instructions.[5] This information is then incorporated in the Supplementary Conditions by amending the appropriate clauses of the General Conditions. A prototype of supplementary conditions subject matter relating to insurance is given among the documents at the end of the chapter.

CERTIFICATES OF INSURANCE

The owner is entitled to evidence of the contractor's insurance coverage in accordance with terms of the General Conditions requiring that certificates

[5] As a guide in this connection the American Institute of Architects has prepared prototype letters covering the request to the owner and the owner's instructions to the architect. See *Architect's Handbook of Professional Practice,* AIA, Chapter 7, Insurance and Bonds of Suretyship (September 1966 Edition), pp. 10, 11, and 12.

acceptable to him be filed before starting work. It is recommended that AIA Document G 705 be used for this purpose. Study of this form, reproduced at the end of the chapter, shows the type of information and amount of detail considered essential. It should be noted that the certificate is to be executed by an authorized representative of the insurance company.

If, on review of the certificate by the owner and his insurance advisor, not all of the required coverage appears not to have been provided, the contractor should be instructed to have the pertinent policies amended and furnish further evidence that all of the required protection has been obtained. No certificate for payment under the construction contract should be issued until the owner informs the architect-engineer that satisfactory evidence of insurance has been furnished by the contractor.

INSURANCE AND BOND CHECK LISTS

This chapter has presented in broad outline the major types of surety bonds and insurance customarily employed in building construction. Many additional forms of protection are available that may be pertinent to the requirements of owner, contractor, or architect-engineer on any particular project. In order to assist in establishing appropriate scope of protection, several professional and trade associations have issued bond and insurance check lists. Among these is the one in the AIA *Architect's Handbook of Professional Practice* which is reproduced at the end of this chapter. Another, pertaining more specifically to requirements of the contractor, is published by the Associated General Contractors of America.

EXAMPLES OF DOCUMENTS

The following documents are reproduced here for illustrative purposes:

Performance and Payment Bonds
Bid Bond
Prototype Supplementary Conditions Relating to Insurance
Certificate of Insurance
Checklist on Insurance and Bond Protection

Reproduction is by permission of the American Institute of Architects.

THE AMERICAN INSTITUTE OF ARCHITECTS

AIA DOCUMENT
SEPT. 1963 ED.   A311

AIA DOC. A311   SEPT. 1963 ED.

PERFORMANCE BOND

KNOW ALL MEN BY THESE PRESENTS: that (Here insert name and address or legal title of Contractor)

as Principal, hereinafter called Contractor, and, (Here insert the legal title and address of Surety)

as Surety, hereinafter called Surety, are held and firmly bound unto (Name and address or legal title of Owner)

as Obligee, hereinafter called Owner, in the amount of

Dollars ($),

for the payment whereof Contractor and Surety bind themselves, their heirs, executors, administrators, successors and assigns, jointly and severally, firmly by these presents.

WHEREAS,

Contractor has by written agreement dated   19   , entered into a contract with Owner for

in accordance with drawings and specifications prepared by (Here insert full name, title and address)

which contract is by reference made a part hereof, and is hereinafter referred to as the Contract.

PERFORMANCE/LABOR-MATERIAL BOND
AIA DOC. A311   SEPT. 1963 ED.   FOUR PAGES

© 1963 The American Institute of Architects   PAGE 1
1735 New York Ave N.W., Washington, D. C.

AIA

NOW,THERFORE, THE CONDITION OF THIS OBLIGATION is such that, if Contractor shall prompltly and faithfully perform said Contract, then this obligation shall be null and void; otherwise it shall remain in full force and effect.

The Surety hereby waives notice of any alteration or extension of time made by the Owner.

Whenever Contractor shall be, and declared by Owner to be in default under the Contract, the Owner having performed Owner's obligations thereunder, the Surety may promptly remedy the default, or shall promptly

1) Complete the Contract in accordance with its terms and conditions, or

2) Obtain a bid or bids for submission to Owner for completing the Contract in accordance with its terms and conditions, and upon determination by Owner and Surety of the lowest responsible bidder, arrange for a contract between such bidder and Owner, and make available as work progresses (even though there should be a default or a succession of defaults under the contract or contracts of completion arranged under this paragraph) sufficient funds to pay the cost of completion less the balance of the contract price; but not exceeding, including other costs and damages for which the Surety may be liable hereunder, the amount set forth in the first paragraph hereof. The term "balance of the contract price," as used in this paragraph, shall mean the total amount payable by Owner to Contractor under the Contract and any amendments thereto, less the amount properly paid by Owner to Contractor.

Any suit under this bond must be instituted before the expiration of two (2) years from the date on which final payment under the contract falls due.

No right of action shall accrue on this bond to or for the use of any person or corporation other than the Owner named herein or the heirs, executors, administrators or successors of Owner.

Signed and sealed this   day of   A.D. 19

IN THE PRESENCE OF:

{
                              (Principal)   (Seal)

_____

                              (Title)

{
                              (Surety)   (Seal)

_____

                              (Title)

THE AMERICAN INSTITUTE OF ARCHITECTS

A311

LABOR AND MATERIAL PAYMENT BOND

THIS BOND IS ISSUED SIMULTANEOUSLY WITH PERFORMANCE BOND IN FAVOR OF THE
OWNER CONDITIONED ON THE FULL AND FAITHFUL PERFORMANCE OF THE CONTRACT

KNOW ALL MEN BY THESE PRESENTS: that (Here insert name and address or legal title of Contractor)

as Principal, hereinafter called Principal, and, (Here insert the legal title and address of Surety)

as Surety, hereinafter called Surety, are held and firmly bound unto (Name and address or legal title of Owner)

as Obligee, hereinafter called Owner, for the use and benefit of claimants as hereinbelow defined, in the

amount of   Dollars ($),
(Here insert a sum equal to at least one-half of the contract price)
for the payment whereof Principal and Surety bind themselves, their heirs, executors, administrators, successors and assigns, jointly and severally, firmly by these presents.

WHEREAS,

Principal has by written agreement dated   19   , entered into a contract with Owner for

in accordance with drawings and specifications prepared by (Here insert full name, title and address)

which contract is by reference made a part hereof, and is hereinafter referred to as the Contract.

PERFORMANCE/LABOR-MATERIAL BOND	FOUR PAGES
AIA DOC. A311 SEPT. 1963 ED.	PAGE 3

NOW, THEREFORE, THE CONDITION OF THIS OBLIGATION is such that, if Principal shall promptly make payment to all claimants as hereinafter defined, for all labor and material used or reasonably required for use in the performance of the Contract, then this obligation shall be void; otherwise it shall remain in full force and effect, subject, however, to the following conditions:

1. A claimant is defined as one having a direct contract with the Principal or with a subcontractor of the Principal for labor, material, or both, used or reasonably required for use in the performance of the contract, labor and material being construed to include that part of water, gas, power, light, heat, oil, gasoline, telephone service or rental of equipment directly applicable to the Contract.

2. The above named Principal and Surety hereby jointly and severally agree with the Owner that every claimant as herein defined, who has not been paid in full before the expiration of a period of ninety (90) days after the date on which the last of such claimant's work or labor was done or performed, or materials were furnished by such claimant, may sue on this bond for the use of such claimant, prosecute the suit to final judgment for such sum or sums as may be justly due claimant, and have execution thereon. The Owner shall not be liable for the payment of any costs or expenses of any such suit.

3. No suit or action shall be commenced hereunder by any claimant:

a) Unless claimant, other than one having a direct contract with the Principal, shall have given written notice to any two of the following: The Principal, the Owner, or the Surety above named, within ninety (90) days after such claimant did or performed the last of the work or labor, or furnished the last of the materials for which said claim is made, stating with substantial accuracy the amount claimed and the name of the party to whom the materials were furnished, or for whom the work or labor was done or performed. Such notice shall be served by mailing the same by registered mail or certified mail, postage prepaid, in an envelope addressed to the Principal, Owner or Surety, at any place where an office is regularly maintained for the transaction of business, or served in any manner in which legal process may be served in the state in which the aforesaid project is located, save that such service need not be made by a public officer.

b) After the expiration of one (1) year following the date on which Principal ceased work on said Contract, it being understood, however, that if any limitation embodied in this bond is prohibited by any law controlling the construction hereof such limitation shall be deemed to be amended so as to be equal to the minimum period of limitation permitted by such law.

c) Other than in a state court of competent jurisdiction in and for the county or other political subdivision of the state in which the project, or any part thereof, is situated, or in the United States District Court for the district in which the project, or any part thereof, is situated, and not elsewhere.

4. The amount of this bond shall be reduced by and to the extent of any payment or payments made in good faith hereunder, inclusive of the payment by Surety of mechanics' liens which may be filed of record against said improvement, whether or not claim for the amount of such lien be presented under and against this bond.

Signed and sealed this   day of   A.D. 19

IN THE PRESENCE OF:

_____

(Principal)   (Seal)

(Title)

_____

(Surety)   (Seal)

(Title)

THE AMERICAN INSTITUTE OF ARCHITECTS

AIA DOCUMENT
SEPT. 1963 ED.   A310

BID BOND

KNOW ALL MEN BY THESE PRESENTS, that we

as Principal, hereinafter called the Principal, and

a corporation duly organized under the laws of the State of
as Surety, hereinafter called the Surety, are held and firmly bound unto

as Obligee, hereinafter called the Obligee, in the sum of

Dollars ($),

for the payment of which sum well and truly to be made, the said Principal and the said Surety, bind ourselves, our heirs, executors, administrators, successors and assigns, jointly and severally, firmly by these presents.

WHEREAS, the Principal has submitted a bid for

NOW THEREFORE, if the Obligee shall accept the bid of the Principal and the Principal shall enter into a contract with the Obligee in accordance with the terms of such bid, and give such bond or bonds as may be specified in the bidding or contract documents with good and sufficient surety for the faithful performance of such contract and for the prompt payment of labor and material furnished in the prosecution thereof, or in the event of the failure of the Principal to enter such contract and give such bond or bonds, if the Principal shall pay to the Obligee the difference not to exceed the penalty hereof between the amount specified in said bid and such larger amount for which the Obligee may in good faith contract with another party to perform the work covered by said bid, then this obligation shall be null and void, otherwise to remain in full force and effect.

Signed and sealed this   day of   A.D. 19

```
                    ┐   _____ (Seal)
_____    ├         Principal
                    ┘   _____
                              Title

                    ┐   _____ (Seal)
_____    ├         Surety
                    ┘   _____
                              Title
```

BID BOND
AIA DOC. A310   SEPT. 1963 ED.   ONE PAGE

PROTOTYPE OF
SUPPLEMENTARY CONDITIONS RELATING TO INSURANCE*

(This hypothetical project involves construction of an addition to an office building including tie-in to the existing building and remodelling the first and second floors, site excavation and underpinning an adjoining building. Contract amount – $2,500,000.)

Paragraph 11.1–Contractor's Liability Insurance
Amplify the paragraph by adding:

a. During the term of the Contract, the Contractor and each Subcontractor shall, at its own expense, purchase and maintain the following insurance in companies properly licensed and satisfactory to Owner.

    1. *Workmen's Compensation including Occupational Disease, and Employer's Liability Insurance*

        a. *Statutory*–Amounts and coverage as required by District of Columbia, Maryland and Virginia Workmen's Compensation laws, including provision for Voluntary D.C. benefits as required in labor union agreements and including the "All States" endorsement.

        b. *Employer's Liability*–At least $500,000 each accident.

    2. *Public Liability*

    including coverage for direct operations, sublet work, elevators, contractual liability and completed operations with limits not less than those stated below.

        a. *Bodily Injury Liability*–including Personal Injuries
          $500,000 each person
          $1,000,000 each occurrence

        b. *Property Damage Liability*
          $500,000 each occurrence
          $500,000 aggregate

    Regarding Property Damage:
    Include Broad Form Property Damage–Remove "XCU" Exclusions (explosion, collapse, underground property damage).
    Regarding Completed Operations Liability:
    Continue coverage in force for one year after completion of the Work.

    3. *Comprehensive Automobile Liability Insurance* including coverage for owned, non-owned and hired vehicles–with limits not less than those stated below.

---

*Reproduced by permission from the *Architect's Handbook of Professional Practice*, AIA, Chapter 7 (September 1966 Edition). Paragraph numbers refer to AIA Document A201 given at the end of Chapter 4.

a. *Bodily Injury Liability*
$500,000 each person
$1,000,000 each occurrence
b. *Property Damage Liability*
$250,000 each occurrence

Liability Insurance may be arranged by Comprehensive General Liability and Comprehensive Automobile Liability policies for the full limits required; or by a combination of underlying Comprehensive Liability policies for lesser limits with the remaining limits provided by an Excess or Umbrella Liability policy.

Paragraph 11.3 Property Insurance

Amplify Subparagraph 11.3.1 to add:— Owner shall increase present Special Multi-Peril Policy for insurable value of the Work and add the names of the Contractor and Subcontractors as interest may appear.

Amplify Subparagraph 11.3.2 to add:— Owner will continue to carry present Steam Boiler Policy with a limit per accident of $250,000 and add the new boiler and the names of the Contractor and Subcontractors as interest may appear.

Bonds

Before commencing work Contractors will furnish two bonds, each in a surety company satisfactory to the Owner; namely, a Performance Bond (AIA Document A311) and a Labor and Material Payment Bond (AIA Document A311). Each Bond shall be in a Penal Sum equal to 100 percent of th e Contract Price.

CERTIFICATE OF INSURANCE
AIA Document G705

This certifies to the Addressee shown below that the following described policies, subject to their terms, conditions and exclusions, have been issued to:
NAME & ADDRESS
OF INSURED

COVERING (Show Project Name
And/Or Number And Location)

Addressee:

Date —————————————

Kind Of Insurance	Policy Number	Inception Date	Expiration Date	Limits Of Liability	
1. (a) Workmen's Comp.				$	Statutory Workmen's Compensation
(b) Employers' Liability				$	One Accident and Aggregate Disease
2. Comprehensive General Liability				$	Each Person—Premises and Operations
				$	Each Person—Elevators
				$	Each Person—Independent Contractors
				$	Each Person— Products Including Completed Operations
(a) Bodily Injury				$	Each Person—Contractual
Including				$	Each Occurrence—
Personal Injury				$	Aggregate Products Including Completed Operations
				$	Each Occurrence—Premises—Operations
				$	Each Occurrence—Elevators
				$	Each Occurrence—Independent Contractors
(b) Property Damage				$	Each Occurrence— Products Including Completed Operations
				$	Each Occurrence—Contractual
				$	Aggregate—
				$	Aggregate— Operations Protective Products and Contractual
3. Comprehensive Automobile Liability				$	Each Person
(a) Bodily Injury				$	Each Occurrence—
(b) Property Damage				$	Each Accident—

4.

UNDER GENERAL LIABILITY POLICY OR POLICIES   Yes   No
1. Does Property Damage Liability Insurance shown include coverage for XC and U hazards?   ____   ____
2. Is Occurrence Basis Coverage provided under Property Damage Liability? . . . .   ____   ____
3. Is Broad Form Property Damage Coverage Provided for this Project? . . . .   ____   ____
4. Is Personal Injury Coverage included? . . . .   ____   ____
5. Is coverage provided for Contractual Liability (Indemnification Clause) assumed by insured?   ____   ____

UNDER AUTOMOBILE LIABILITY POLICY OR POLICIES
1. Does coverage shown above apply to non-owned and hired automobiles?   . . . .   ____   ____
2. Is Occurrence Basis Coverage provided under Property Damage Liability? . . . .   ____   ____

In the event of cancellation, fifteen (15) days
written notice shall be given to the party to whom   _____
this Certificate is addressed.   NAME OF INSURANCE COMPANY

                                                        _____
                                                        ADDRESS

                                                        _____
                                                        SIGNATURE OF AUTHORIZED REPRESENTATIVE

This document is copyrighted by the American Institute of Architects and is reproduced here with its permission.

CHECKLIST ON PROTECTION*

This checklist is based upon the foregoing discussion of the more common risks and insurance or bond coverages to be considered by architects, owners, contractors and their attorneys and insurance counselors.

Architect's Liability—
1. Professional Liability including Contractual ("hold harmless")
2. Comprehensive Personal Liability
3. Comprehensive General Liability —Occurrence Basis
   a. Premises—Operations
   b. Elevator
   c. Contingent
   d. Contractural
   e. Completed Operations—Products
   f. Broad Form Property Damage Endorsement
   g. Property Damage XCU Endorsement (Explosion, Collapse, and Underground Damage)
   h. Personal Injury Endorsement (Invasion of Privacy, False Arrest, Libel, Slander, Defamation of Character)
4. Excess or Umbrella Liability
5. Comprehensive Automobile Liability—Occurrence Basis
6. Aircraft or Watercraft Liability
7. Fire Legal Liability
8. Water Damage and/or Sprinkler Leakage Legal Liability
9. Nuclear Energy Liability

Architect's Personnel—
1. Workmen's Compensation & Employers Liability
2. Disability Income—Salary Continuance
3. Major Medical
4. Hospitalization—Surgical Expense
5. Life Insurance—Group-Key Man-Partnership
6. Accident Insurance—Death, Permanent Disability
7. Retirement—Pension—Deferred Compensation

Architect's Office—
1. All Physical Loss—Building or Leasehold Improvements
2. Fire, Extended Coverage and Vandalism
3. Boiler and Machinery
4. Water Damage—Sprinkler Leakage
5. Collapse
6. Comprehensive Glass
7. Demolition Endorsement
8. Office Contents Special ("All Risk") Form
9. Valuable Documents
10. Equipment Floater
11. Automobile Material Damage & Collision
12. Business Interruption—Loss of Use
    a. Business Interruption—Fire, Extended Coverage or All Physical Loss
    b. Rental Value—Fire, Extended Coverage or All Physical Loss
    c. Extra Expense—Fire, Extended Coverage or All Physical Loss
    d. Boiler & Machinery Use and Occupancy
    e. Water Damage—Sprinkler Leakage Use and Occupancy
    f. Leasehold Interest
13. Package Policy
14. Theft-Robbery-Burglary
15. Broad Form Money & Securities
16. Fidelity Bonds
17. Forgery Bonds
18. Credit Card Forgery
19. Blanket Crime Policy—Comprehensive Dishonesty, Disappearance, Destruction Bond

This checklist should be applied to each project in accordance with the prevailing circumstances and requirements. In all insurance matters the determination of specific coverages is the function of insurance counselors; it is not the responsibility of the architect.

*Reproduced by permission from the *Architect's Handbook of Professional Practice,* AIA, Chapter 7 (September 1966 Edition).

This checklist is not represented as being all-embracing and the practioner should consult his insurance counselor to determine his needs.

Contractor Requirements—

1. Workmen's Compensation & Employers Liability
2. Comprehensive General Liability —Occurrence Basis
   a. Premises—Operations
   b. Elevator—Hoists
   c. Contingent
   d. Contractural
   e. Completed Operations—Products
   f. Broad Form Property Damage Endorsement
   g. Property Damage XCU Endorsement (Explosion, Collapse, Underpinning)
   h. Personal Injury Endorsement
3. Comprehensive Automobile Liability—Occurrence Basis
4. Aircraft or Watercraft Liability
5. Nuclear Energy Liability
6. Excess or Umbrella Liability
7. Contractor's Equipment Floater
8. Workmen's Tools Floater
9. Transportation Floater
10. Installation Floater
11. Automobile Material Damage & Collision
12. Broad Form Money & Securities
13. Fidelity Bonds
14. Forgery Bonds
15. Theft-Robbery-Burglary
16. Blanket Crime Policy—Comprehensive Dishonesty, Disappearance, Destruction Bond
17. Surety Bonds—See Owner Requirements
18. Property Insurance—Fire, Extended Coverage and other perils.

Owner Requirements—

1. Workmen's Compensation & Employers Liability
2. Comprehensive General Liability —Occurrence Basis
   a. Premises—Operations
   b. Elevator
   c. Contingent
   d. Contractual
   e. Completed Operations—Products
   f. Broad Form Property Damage Endorsement
   g. Property Damage XCU Endorsement
   h. Personal Injury Endorsement
3. Comprehensive Automobile Liability—Occurrence Basis
4. Aircraft or Watercraft Liability
5. Nuclear Energy Liability
6. Excess or Umbrella Liability
7. Property Insurance
   a. All Physical Loss—Builder's Risk Completed Value Form or Reporting Form
   b. Fire, Extended Coverage & Vandalism—Builder's Risk Completed Value Form or Reporting Form
   c. Boiler and Machinery
   d. Comprehensive Glass
   e. Electric or Neon Sign Form
   f. Bridge-Tunnel-Radio or TV Towers—Special Forms
   g. Water Damage—Sprinkler Leakage
   h. Collapse
   i. Nuclear Energy Property Insurance
   j. Radioactive Contamination
   k. Demolition
8. Business Interruption—Loss of Use
   a. Business Interruption—(Fire, Extended Coverage or All Physical Loss)
   b. Rental Value (Fire, Extended Coverage or All Physical Loss)
   c. Extra Expense (Fire, Extended Coverage or All Physical Loss)
   d. Boiler and Machinery Use & Occupancy
   e. Water Damage—Sprinkler Leakage Use & Occupancy
9. Legal Liability for Property of Others in Insured's Custody
   a. Warehousemen
   b. Safe Depository
   c. Fur Storage
   d. Cleaners & Laundries
   e. Garage Keepers

f. Other Bailees
10. Package Policy
11. Surety Bonds
  a. Bid
  b. Performance
  c. Labor and Material Payment
  d. License or Permit
  e. Lien
  f. No Lien Bond
  g. Maintenance
  h. Release of Retained Percentage

  i. Statutory
  j. Subcontract
12. Automobile Material Damage & Collision
13. Broad Form Money & Securities
14. Fidelity Bonds
15. Forgery Bonds
16. Theft-Robbery-Burglary
17. Blanket Crime Policy—Comprehensive Dishonesty, Disappearance, Destruction Bond

Chapter 7

Bidding and Award Procedures

Nature of the Bidding Process

The distinguishing feature of the competitive bidding process is the requirement that sealed offers to build the project for a specific sum be submitted by each bidder at a specified time and place. To implement the process the owner solicits proposals from prospective bidders by issuing invitations to bid or, for public work, by advertising for them. Those indicating their desire to bid on the project are then issued the necessary construction documents together with additional information and instructions to facilitate preparation of bids on the special proposal forms required. At the place and hour specified the sealed bids are opened and tabulated and then examined by the architect-engineer for analysis and comparison. Subsequently, if the lowest acceptable bid is within the range of funds available, a construction contract is awarded. On its execution the successful bidder becomes the contractor for the project.

This chapter deals with the special documents used in the bidding process, the procedures controlling bidding and contract award, and factors within the building industry that exert an influence on effective operation of the competitive bidding system.

Bidding Documents

These consist of the following three items under the heading "Bidding Requirements" in the typical table of contents for the Project Manual or Construction Documents Book discussed at the beginning of Chapter 5.

214

Invitation to bid (or advertisement)
Instructions to bidders
Proposal form

Although the instructions to bidders document is customarily bound into the Project Manual, practice varies with respect to the other two items. This is no doubt because the invitation (or in the case of public work, the advertisement) is circulated ahead of the time prospective bidders see the Project Manual, and the proposal form is a document that is executed by a bidder and stands independently as an official offer. Nevertheless many architects and engineers bind a copy of both these documents into the Project Manual for information and reference purposes.

BID SECURITY

It will be noted that a bid bond is not included among the bidding documents listed. The purpose of such a bond in underwriting the integrity of a bid was discussed in Chapter 6, but the same result can also be achieved by requiring a certified check to accompany the proposal. The amount of the bid security whether required as bond or check may be called for as a percentage of the bid or as a flat sum. The custom of stipulating that the dollar amount of the security ". . . shall be equal to 10% of the bid price" is one of long standing. It has, however, certain disadvantages in that the bond or check cannot be written until the actual amount of the bid is known (frequently only a few hours before the scheduled opening) and that the process of procuring bid bond or certified check in effect reveals the amount of the bid. To ameliorate this situation the amount of bid security may be called for as a flat sum determined by the architect and owner in advance of calling for bids. This can be based on a percentage of the architect's estimate of the cost of the project, but the percentage used need not be revealed. This procedure is recommended for all private work and for public work where controlling statutes do not prescribe otherwise.

The choice between bid bonds and certified checks as bid security (sometimes called bid deposit) is up to the owner and architect-engineer insofar as private construction is concerned, but may be controlled by statute in public work. From the bidder's point of view certified checks immobilize operating funds and this may become a serious consideration when he is bidding on several jobs at about the same time. On the other hand, the ability to furnish a certified check on a large project indicates some measure of the financial resources of the contracting firms bidding, an assurance not necessarily provided by the ability of a bidder to procure a bid bond. Whichever form of bid security is to be used, stipulations concerning it should appear in the invitation or advertisement.

SPECIAL BIDDING DOCUMENTS

The documents discussed in this chapter are those customarily used for bidding private work. Although examples are given of forms employed in public construction, additional documents may be required for bidding on projects sponsored by governmental agencies at federal, state, and local levels. An example of the latter is the bidder's certificate regarding equal opportunity employment called for on federally assisted projects.[1] Many public agencies also give very explicit instructions concerning preparation of bids; however, detailed discussion of such procedures is not presented in this book.

Invitation to Bid

On private work the purpose of the invitation is to establish a list of bidders acceptable to the owner and architect-engineer. In general, a list of six bidders of comparable qualifications and reputation should be sufficient to assure adequate price competition if each firm on the list submits a serious proposal. Since some contractors who are not in a position to take on additional work are reluctant to decline to bid after being invited, they sometimes submit a "courtesy bid" with a price sufficiently high so that there is no chance of its becoming the low bid. Such a bid is, of course, no "courtesy" to owner or architect since it undermines the basis of the competitive bidding system. To provide against this contingency the number of firms to whom invitations are sent is often expanded, but it should be borne in mind that too long a list may discourage close figuring since the chance of any particular bidder coming in low decreases as the number of those submitting proposals increases. This will, of course, be influenced by many other factors that contribute to the "bidding climate," as discussed later in the chapter. Nevertheless, it is worth making the point here that a long bidding list made up of firms possessing comparable degrees of skill, integrity, and responsibility may not assure the lowest price, while appreciable differences in qualifications and reputation of those on the list will encourage courtesy bids by the better firms and may lead to an award that jeopardizes the intended quality of the completed project.

    The formality of the invitation to bid varies with the type and scope of a particular project and the relationships existing between architect-engineer and the firms to be invited. If all contractors on the list have worked with the A & E office before, and if the list is sufficiently limited, the invitation may take the form of a letter. When the project is more complex and nonlocal contractors who have not worked with the architect previously are to be invited, a more

---

[1] See "Certification of Bidder Regarding Equal Employment Opportunity," form HUD-4238-CD-1, U.S. Department of Housing and Urban Development.

formal document is indicated. In any event, the information to be presented in the invitation to bid should include the following.

1. Name and location of the project.
2. Brief description of the work.
3. Type of award, that is, single or separate contracts.
4. Time and place for receiving bids.
5. Statement as to whether opening of bids will be public or private.
6. Location where documents may be examined and procured for bidding purposes.
7. Bid security requirements.
8. Statement relating to owner's right of acceptance or rejection of bids.

These items present information needed by a prospective bidder to decide whether he wishes to pursue the matter further by asking for the drawings and other construction documents necessary to prepare a bid.

DOCUMENT DEPOSITS

It should be noted with respect to item 5 that a deposit is usually required to obtain sets of the plans and other construction documents for bidding purposes. The invitation to bid should state the amount of this deposit and the conditions under which it will be returned after the opening of bids. On large projects where several bidders have been invited it may be desirable to file plans, specifications, etc., for examination at locations other than the architect's office. The "plan rooms" established by various building trade and marketing organizations are appropriate for this purpose as well as for filing additional sets to facilitate access to the construction documents by subcontracting firms during the bidding period. Normally the procurement of sets of construction documents by prime bidders and the attendant receipt of document deposits is handled directly by the architect's office.

Advertisement for Bids

On public construction projects competitive bidding has a threefold objective: to stimulate competition, prevent favoritism and fraud in the award of contracts, and to give an equal opportunity to all those desiring to do business with the public. The laws controlling public construction usually require that an announcement concerning bidding for a project be published in one or more designated newspapers or trade magazines. This announcement constitutes the advertisement or legal notice to bidders. In some jurisdictions the form and content of the advertisement are specified by statute or municipal charter, but the information presented must cover substantially the same items as those listed in the preceding article for an invitation to bid. In addition, advertisements usually contain a statement indicating that the successful bidder will be required to furnish performance and payment bonds. Furthermore, reference may be made to bids being received only from bidders who have complied with prequalification requirements announced by legal notice published several days before the advertisement. The efficacy of prequalification procedures in both public and private construction is discussed later in this chapter. An example of a public[2] advertisement for bids under the separate contract system follows.

NEW YORK STATE
ALBANY SOUTH MALL PROJECT

NOTICE TO BIDDERS

Separate sealed proposals covering Construction, Heating, Sanitary, Electric, Elevator Work and Moving Stairs for Legislative Building Superstructure, Albany, New York, Albany South Mall Project, in, accordance with Specifications Nos. 19608-C, 19608-H, 19608-S, 19608-E, 19608-EL, 19608-MS and accompanying drawings, will be received by Director, Contracts Unit, Department of Public Works, Administration and Engineering Building, 1220 Washington Ave., State Campus, Albany, N.Y. 12226, until 10:30 A.M., on Wednesday, August 9, 1967, when they will be publicly opened and read.

Each proposal must be made upon the form and submitted in the envelope provided therefor and shall be accompanied by a certified check made payable to the New York State Department of Public Works, in the amount stipulated in the proposal as a guaranty that the bidder will enter into the contract if it be awarded to him. The specification number must be written on the front of the envelope. The blank spaces in the proposal must be filled in, and no change shall be made in the phraseology of the proposal. The State reverses the right to reject any or all bids. Successful bidders will be required to give a bond conditioned for the faithful performance of the contract and a separate bond for the payment of laborers and materialmen, each bond in the sum of 100% of the amount of the contract.

Drawings and specifications may be examined free of charge at the following offices:

State Architect, 270 Broadway, New York City
State Architect, Division of Architecture Building, State Campus, Albany, N.Y.
District Supervisor of Bldg. Constr., State Office Building, 333 E. Washington St., Syracuse, N.Y.
District Supervisor of Bldg. Constr., Genesee Valley Regional Market, 900 Jefferson Rd., Rochester, N.Y.

---

[2]From *Engineering News-Record*, July 6, 1967, p. 71.

District Engineer, 125 Main St., Buffalo, N.Y.

Drawings and specifications may be obtained by calling at the Contracts Unit, Department of Public Works, Administration and Engineering Building, 1220 Washington Avenue, State Campus, Albany, N.Y., 12226, or at the State Architect's Office, 18th Floor, 270 Broadway, New York City and by making deposit of $200.00 for all related trades on this project; or as follows for each separate trade: Construction $100.00; Heating, $50.00; Sanitary, $50.00; Electric, $50.00; Elevator, $40.00; Moving Stairs, $20.00; or by mailing such deposit to the Albany address. Checks should be made payable to the State Department of Public Works. Proposal blanks and envelopes will be furnished without charge. The State Architect's Standard Specifications of January 2, 1960 will be required for this project and may be purchased from the Bureau of Fiscal Administration, Department of Public Works, Administration and Engineering Building, State Campus, Albany, N.Y., or at the office of the State Architect, 270 Broadway, New York City, or at the offices of District Supervisor of Bldg. Constr., State Office Building, 333 E. Washington St., Syracuse, N.Y., or District Supervisor of Bldg. Constr., Genesee Valley Regional Market, 900 Jefferson Road, Rochester, N.Y., for the sum of $5.00 each.

The completion date for this project is June 30, 1970.

Instructions to Bidders

This document, also frequently called information for bidders, contains detailed requirements pertaining to the submission of bids. Although the character of the information is similar to that contained in the invitation or advertisement, it is expanded and presented in more detail with a view to assisting a bidder in the preparation of his proposal. It is common practice in private work to compose an individual set of instructions for each project, although many of the items covered may represent more or less standard procedure in the architect's office concerned. On the other hand, public construction agencies frequently use a printed form covering standardized policy requirements and then state the details for a particular project in a supplement.

A question arises whether to duplicate information in the instructions to bidders that is given in the invitation to bid. Practice in this regard varies. Some architects and engineers feel that the instructions should be self-contained and consider project data presented in the invitation as an abstract or summary of the more complete information in the instructions. Others contend that once something has been said in the invitation it should not be repeated and should be cited in the instructions by reference only. The comments which follow are predicated on the instructions to bidders constituting a self-contained document. Consequently, in addition to other information, each item mentioned in the invitation should have its counterpart restated with or without amplification as circumstances may direct.

IDENTIFICATION AND DESCRIPTION OF PROJECT

This should contain the official name and location of the work, the name and address of the owner (or awarding authority if different), a brief general description of the type and size of the project, the time allowed for completion, and provision for assessing liquidated damages, if any.

TYPE OF AWARD

If construction will be accomplished under the separate contracts system, arrangements for coordinating the work of the separate contractors should be described and the responsibility therefor should be designated.

RECEIVING AND OPENING BIDS

Under this heading should appear the person, office, or agency that will receive bids, the place and time for receiving them, and the location and time of the bid opening. It should be stated whether the opening will be public or private and if private, whether a representative of each prime bidder may attend.

DOCUMENTS

This item should give the detailed information concerning procurement and availability of documents for bidding purposes. It may be substantially a repetition of similar information in the invitation to bid, although space limitations in that document often preclude detailed statements pertaining to items such as whether plans and specifications may be obtained by other than prime bidders, provisions for issuing partial sets, and procedure for refund of deposits.

CONDITIONS AFFECTING THE WORK

Examination of the site is required in order to ascertain whether local conditions or other factors peculiar to the location can affect the cost of the work.

CLARIFICATION OF DOCUMENTS

A procedure must be prescribed for handling inquiries concerning apparent discrepancies and ambiguities in the plans and specifications and questions about interpretation of their provisions. Usually, only written inquiries are accepted and replies are issued as addenda sent to all bidders of record. Obviously, a cutoff date for accepting inquiries must be established in order that all bidders may receive the last addendum a few days before the bid opening date.

SUBSTITUTION OF MATERIALS

When substitutions are to be considered during the bidding period, it is necessary to establish a procedure for handling such requests. This matter is discussed in Chapter 5 in connection with types of specifications. When requested substitutions are found acceptable, this information is made known to all bidders of record through issue of an addendum. As with document clarification requests, some cutoff date for receipt of substitution requests must be established sufficiently in advance of bid opening to permit proper processing.

PREPARATION OF BID

Additional information should be given to assist in preparing and submitting the bid on the prescribed proposal form. This includes stating the number of copies required (usually two or three), explicit instructions concerning signatures, requirements for submitting bid security with the proposal, information to appear on the envelope in which the proposal and bid security are sealed, and other special instructions.

BID SECURITY REQUIREMENTS

Under this item information given in the invitation to bid should be expanded to specify the exact title of the obligee where bid bonds are called for or the individual or agency in whose favor a certified check should be drawn. In addition, the length of time that bid securities will be retained should be stated and the mechanism for their return to unsuccessful bidders described.

LISTING OF SUBCONTRACTORS

When such procedure is required, as under certain regulations of the General Services Administration and also often in private work, the extent of a bidder's obligation in this regard should be clearly stated. Paragraphs similar to the one following (either with or without the provision for multiple names) are sometimes used for this purpose in private construction:

> *The bidders shall each list in the space provided on the proposal form, the names and addresses of all subcontractors proposed for all divisions of work listed therein. In order not to restrict the operations of the contractor, bidders are granted the right to submit up to three names for each division of the work if they desire.*

It should be noted that the listing of subcontractors at time of submitting proposals is by no means a universal practice.[3]

MODIFICATION OF PROPOSAL

It is customary to permit modification or withdrawal of proposals after submittal but before the bid opening; where this is allowed the procedure should be specified, including any time limits that may apply.

---

[3]For other methods of controlling the selection of subcontractors see AIA Document A201, Article 5, reproduced at the end of Chapter 4.

OWNER'S RIGHT OF REJECTION

Circumstances that may lead to disqualification of a bid should be described, and reservation of the owner's right to waive irregularities in proposals or to reject any or all bids should be stated.

EXAMPLE OF INSTRUCTIONS

It was noted earlier that in private work it is customary to prepare an individual set of instructions to bidders for each project. However, many A & E firms pattern the clauses of such documents on those contained in Standard Form 22 of the General Services Administration reproduced here. Examination of this form shows that some items obviously apply to federal government work only, whereas a few of those discussed above relating primarily to private work do not appear at all. Nevertheless, the main provisions covered in the foregoing discussion and in Form 22 are substantially similar.

STANDARD FORM 22
JUNE 1964 EDITION
GENERAL SERVICES ADMINISTRATION
FED. PROC. REG. (41 CFR) 1-16.401

INSTRUCTIONS TO BIDDERS
(CONSTRUCTION CONTRACT)

1. Explanations to Bidders. Any explanation desired by a bidder regarding the meaning or interpretation of the invitation for bids, drawings, specifications, etc., must be requested in writing and with sufficient time allowed for a reply to reach bidders before the submission of their bids. Any interpretation made will be in the form of an amendment of the invitation for bids, drawings, specifications, etc., and will be furnished to all prospective bidders. Its receipt by the bidder must be acknowledged in the space provided on the Bid Form (Standard Form 21) or by letter or telegram received before the time set for opening of bids. Oral explanations or instructions given before the award of the contract will not be binding.

2. Conditions Affecting the Work. Bidders should visit the site and take such other steps as may be reasonably necessary to ascertain the nature and location of the work, and the general and local conditions which can affect the work or the cost thereof. Failure to do so will not relieve bidders from responsibility for estimating properly the difficulty or cost of successfully performing the work. The Government will assume no responsibility for any understanding or representations concerning conditions made by any of its officers or agents prior to the execution of the contract, unless included in the invitation for bids, the specifications, or related documents.

3. Bidder's Qualifications. Before a bid is considered for award, the bidder may be requested by the Government to submit a statement regarding his previous experience in performing comparable work, his business and technical organization, financial resources, and plant available to be used in performing the work.

4. Bid Guarantee. Where a bid guarantee is required by the invitation for bids, failure to furnish a bid guarantee in the proper form and amount, by the time set for opening of bids, may be cause for rejection of the bid.

A bid guarantee shall be in the form of a firm commitment, such as a bid bond, postal money order, certified check, cashier's check, irrevocable letter of credit or, in accordance with Treasury Department regulations, certain bonds or notes of the United States. Bid guarantees, other than bid bonds, will be returned (a) to unsuccessful bidders as soon as practicable after the opening of bids, and (b) to the successful bidder upon execution of such further contractual documents and bonds as may be required by the bid as accepted.

If the successful bidder, upon acceptance of his bid by the Government within the period specified therein for acceptance (sixty days if no period is specified) fails to execute such further contractual documents, if any, and give such bond(s) as may be required by the terms of the bid as accepted within the time specified (ten days if no period is specified) after receipt of the forms by him, his contract may be terminated for default. In such event he shall be liable for any cost of procuring the work which exceeds the amount of his bid, and the bid guarantee shall be available toward offsetting such difference.

5. Preparation of Bids. (a) Bids shall be submitted on the forms furnished, or copies thereof, and must be manually signed. If erasures or other changes appear on the forms, each erasure or change must be initialed by the person signing the bid. Unless specifically authorized in the invitation for bids, telegraphic bids will not be considered.

(b) The bid form may provide for submission of a price or prices for one or more items, which may be lump sum bids, alternate prices, scheduled items resulting in a bid on a unit of construction or a combination thereof, etc. Where the bid form explicitly requires that the bidder bid on all items, failure to do so will disqualify the bid. When submission of a price on all items is not required, bidders should insert the words "no bid" in the space provided for any item on which no price is submitted.

(c) Unless called for, alternate bids will not be considered.

(d) Modifications of bids already submitted will be considered if received at the office designated in the invitation for bids by the time set for opening of bids. Telegraphic modifications will be considered, but should not reveal the amount of the original or revised bid.

6. Submission of Bids. Bids must be sealed, marked, and addressed as directed in the invitation for bids. Failure to do so may result in a premature opening of, or a failure to open, such bid.

7. Late Bids and Modifications or Withdrawals. (a) Bids and modifications or withdrawals thereof received at the office designated in the invitation for bids after the exact time set for opening of bids will not be considered unless: (1) They are received before award is made; and either (2) they are sent by registered mail or by certified mail for which an official dated post office stamp (postmark) on the original Receipt for Certified Mail has been obtained, or by telegraph if authorized, and it is determined by the Government that the late receipt was due solely to delay in the mails, or delay by the telegraph company, for which the bidder was not responsible; or (3) if submitted by mail (or by telegram if authorized), it is determined by the Government that the late receipt was due solely to mishandling by the Government after receipt at the Government installation: *Provided,* That timely receipt at such installation is established upon examination of an appropriate date or time stamp (if any) of such installation, or of other documentary evidence of receipt (if readily available) within the control of such installation or of the post office serving it. However, a modification which makes the terms of the otherwise successful bid more favorable to the Government will be considered at any time it is received and may thereafter be accepted.

(b) Bidders using certified mail are cautioned to obtain a Receipt for Certified Mail showing a legible, dated postmark and to retain such receipt against the chance that it will be required as evidence that a late bid was timely mailed.

(c) The time of mailing of late bids submitted by registered or certified mail shall be deemed to be the last minute of the date shown in the postmark on the registered mail receipt or registered mail wrapper or on the Receipt for Certified Mail unless the bidder furnishes evidence from the post office station of mailing which establishes an earlier time. In the case of certified mail, the only acceptable evidence is as follows: (1) where the Receipt for Certified Mail identifies the post office station of mailing, evidence furnished by the bidder which establishes that the business day of that station ended at an earlier time, in which case the time of mailing shall be deemed to be the last minute of the business day of that station; or (2) an entry in ink on the Receipt for Certified Mail showing the time of mailing and the initials of the postal employee receiving the item and making the entry, with appropriate written verification of such entry from the post office station of mailing, in which case the time of mailing shall be the time shown in the entry. If the postmark on the original Receipt for Certified Mail does not show a date, the bid shall not be considered.

8. Withdrawal of Bids. Bids may be withdrawn by written or telegraphic request received from bidders prior to the time set for opening of bids.

9. Public Opening of Bids. Bids will be publicly opened at the time set of opening in the invitation for bids. Their content will be made public for the information of bidders and others interested, who may be present either in person or by representative.

10. Award of Contract. (a) Award of contract will be made to that responsible bidder whose bid, conforming to the invitation for bids, is most advantageous to the Government, price and other factors considered.

(b) The Government may, when in its interest, reject any or all bids or waive any informality in bids received.

(c) The Government may accept any item or combination of items of a bid, unless precluded by the invitation for bids or the bidder includes in his bid a restrictive limitation.

11. Contract and Bonds. The bidder whose bid is accepted will, within the time established in the bid, enter into a written contract with the Government and, if required, furnish performance and payment bonds on Government standard forms in the amounts indicated in the invitation for bids or the specifications.

U.S. GOVERNMENT PRINTING OFFICE: 1964–O–739–186

Proposal Form

The proposal form, also known as the bid form, probably varies more from project to project than either of the other two bidding documents. A special form should be prepared by the architect-engineer for each project so that the required information submitted by each bidder will appear in the same relative position in all proposals. This is essential, particularly when several bid items are involved, to facilitate comparison of bids and to detect omissions or other irregularities.

Since the executed proposal constitutes an offer on the part of the bidder, it is addressed to the owner, usually in care of the architect on private construction work. In addition to specific identification of the project there are four principle elements to be included on the proposal form: the bid items, the pledges, acknowledgments, and reference to enclosure of bid security. The order in which these appear on the bid form will vary according to preference or custom of individual architect-engineers and government awarding authorities.

BID ITEMS

These, of course, are the focal point of the proposal. In the simplest case there is only one item to bid on, the lump-sum price for constructing the project in accordance with the Contract Documents. Occasionally, where the element of time is of great importance two bid items are listed, both reading as above but with Item No. 1 calling for a completion date based on a normal construction schedule and Item No. 2 an earlier completion date based on an accelerated schedule.

When there is some doubt about the possibility of receiving a low bid for the entire project within the funds available, alternate bid items may be incorporated. These should be used with discretion, since a large number of "alternates" increases the complexity of bidding and may introduce sufficient uncertainties with respect to their effect on each other to raise the level of bids received. Under tight budget conditions many architects and engineers attempt to adjust the scope and material quality of a project to assure an acceptable "base bid" for an adequate facility and then provide for additive alternates which may be accepted if funds are available. Under these conditions the Base Bid item would again read as indicated, and Alternate A might call for the additional lump-sum price to furnish and install "Elevators Nos. 3 and 4 in accordance with Alternative Specification A." Alternate B might request a price for substituting a high quality vinyl floor tile for the less expensive floor finish which was originally specified. Of course, "deductive" alternates may also be called for, but there is some question whether they are as effective in the effort to achieve the most value within the construction budget available.

Another bid item that is seen frequently concerns unit prices. In building construction where foundation conditions may be uncertain at the time of bidding, unit prices are often called for to cover cost per lineal foot of additional piling that may be required, cost per cubic yard of additional foundation concrete, etc. This procedure may obviously be extended to any building components that are indefinite at the time of bidding. However, extended use of unit-price proposal items is much more common in heavy civil engineering construction than in building projects.

PLEDGES

These consist of statements so phrased that the bidder, in submitting his proposal, pledges himself to:

(1) execute a contract with the owner and furnish performance and payment bonds, provided he receives written notice of acceptance of his bid within a stipulated number of days after the opening of bids;

(2) commence work within a stipulated number of days after execution of the contract;

(3) complete the work within the time limit established in the Supplementary Conditions;

(4) accomplish the work in accordance with the Contract Documents.

ACKNOWLEDGMENTS

The principal item under this heading is acknowledgment of receipt of addenda issued during the bidding period. It is customary to provide space for the bidder to enter the identifying number of each addendum received, and a statement is often included whereby the bidder indicates that all work covered by the addenda is included in the proposal. It is also customary to obtain acknowledgment that the bidder has examined the Drawings, Specifications, and other Contract Documents and has examined conditions at the site that might affect the work.

BID SECURITY

Because of the importance of bid security to the functioning of the bidding process, one of the items on the proposal form refers specifically to the fact that the required ". . . bid security, consisting of . . . is enclosed."

OTHER ENCLOSURES

Where additional requirements are imposed such as furnishing a list of subcontractors or certification of compliance with special regulations, a statement that these items are enclosed should also appear on the bid form.

EXAMPLES

Although there is wide variation in the format used for proposals, the accompanying reproduction of Standard Form 21 of the General Services Administration illustrates the wording of clauses and general arrangement as encountered in practice. Here again, some items on this form will be seen to apply to federal construction projects only, but the detailed information called for in connection with bidder's signature is commonly required also in private work.

STANDARD FORM 21	1. REFERENCE
DECEMBER 1965 EDITION	APPROP. NO.
GENERAL SERVICES ADMINISTRATION	PROJECT NO.
FED. PROC. REG. (41 CFR)1–16.401	CONTRACT NO.

BID FORM

(CONSTRUCTION CONTRACT)

Read the Instructions to Bidders (Standard Form 22) 3. DATE OF INVITATION

2. This form to be submitted in ☐ DUPLICATE ☐ TRIPLICATE

4. NAME AND LOCATION OF PROJECT   5. NAME OF BIDDER
                                            (Type or print)

☐ CONSTRUCTION ☐ EXTENSION & REMODELING ☐ REPAIR & IMPROVEMENT

6. TO: GENERAL SERVICES ADMINISTRATION
       PUBLIC BUILDINGS SERVICE

                                        _____
                                                 (Date)

In compliance with the above-dated invitation for bids, the undersigned hereby proposes to perform all work for

in struct accordance with the General Provisions (Standard Form 23–A), Labor Standards Provisions Applicable to Contracts in Excess of $2,000 (Standard Form 19–A), specifications, schedules, drawings, and conditions, for the following amount(s)

_____ Dollars ($_____)

21-108-03   (Continue on other side)   (GSA OVERPRINT JUN 1966)

The undersigned agrees that, upon written acceptance of this bid, mailed or otherwise furnished within   calendar days (calendar days unless a different period be inserted by the bidder) after the date of opening of bids, he will within 15 calendar days (unless a longer period is allowed) after receipt of the prescribed forms, execute Standard Form 23, Construction Contract, and give performance and payment bonds on Government standard forms with good and sufficient surety.

The undersigned agrees, if awarded the contract, to commence the work as soon as practicable after the date of receipt of notice to proceed, and to complete the work within the number of calendar days after the date of receipt of notice to proceed as stipulated in the SPECIAL CONDITIONS.

RECEIPT OF AMENDMENTS: The undersigned acknowledges receipt of the following amendments of the invitation for bids, drawings, and/or specifications, etc. (Give number and date of each):

AMENDMENT NO.						
DATE						
AMENDMENT NO.						
DATE						

The representations and certifications on the accompanying STANDARD FORM 19–B are made a part of this bid.

ENCLOSED IS BID GUARANTEE CONSISTING OF	IN THE AMOUNT OF
NAME OF BIDDER (Type or print)	FULL NAME OF ALL PARTNERS (Type or print)
BUSINESS ADDRESS (Type or print) (Include "ZIP Code")	
BY (Signature in ink. Type or print name under signature)	
TITLE (Type or print)	

DIRECTIONS FOR SUBMITTING BIDS: Envelopes containing bids, guarantee, etc., must be sealed, marked, and addressed as follows:

Mark envelope in lower left corner: 7. PROJECT NO. 8. TO BE OPENED:	9. A D D R E S S	General Services Administration Public Buildings Service, Region Room No.

10.

CAUTION—Bids should not be qualified by exceptions to the bidding conditions.

The Bid Opening

This, of course, is an event of considerable significance to all concerned. By the time this "ceremony" is over, the owner will know whether his budget is sufficient to permit construction of the project; the architect-engineer can either congratulate himself on the accuracy of his prebidding estimate or deplore its inadequacy; and the low bidder can rejoice in his apparent success in acquiring new business or wonder whether the rather substantial difference between his bid and the next higher one indicates a mistake in his figuring.

The nature of the process of bid preparation is such that careful attention to the day named for the opening is very worthwhile. The preferred timing is afternoon on a Tuesday, Wednesday, Thursday, or Friday, but not on a legal holiday or the day following.[4] It is also desirable that there be cooperation among A & E offices to avoid, insofar as practicable, conflict of bid opening dates for important projects in the area.

The degree of formality with which openings are conducted varies from the most scrupulous observance of mandatory procedures in connection with public construction to a more casual atmosphere in much private work. In all public openings and in private openings where a representative of each bidder is invited, the principal steps in the procedure consist of breaking the sealed envelope containing a bidder's proposal, calling out the bidder's acknowledgment of receipt of addenda and the type and amount of bid security enclosed, and reading aloud the amount of each bid item. This process is then repeated for each proposal received. Many architects and engineers prepare bid tabulation forms, which are distributed at the opening for the convenience of those desiring to make a record of the bidding. When bids are not opened in the presence of the bidders, a tabulation of all bids received should be furnished each bidder.

The bid opening is usually adjourned without any official statement being made concerning who is the low bidder, but those present can, of course, draw their own conclusions. Before an acceptable low bidder can be designated, a careful analysis and comparison of bids is made by the architect-engineer and a conference may be held with the apparent low bidder. This is known as "canvassing the bids," and it is discussed in more detail in the following article.

Canvassing the Bids

The bidder's pledge to enter into a contract if his proposal is accepted within a stipulated number of days following the bid opening creates an interval during which all the offers remain open. This period (usually from ten to thirty days) is

---

[4] This timing is recommended by both the American Institute of Architects and the Associated General Contractors of America.

used for canvassing the bids and gives the owner an opportunity to accept the proposal of another bidder without repeating the bidding process in event the low bidder defaults or is disqualified for any reason. It is customary, however, to return the bid security of those at the high end of the list within a few days, especially when this has been furnished by certified check.

MISTAKES IN BIDS

Discussion with the low bidder assumes special importance when his bid appears to be out of line in comparison with the others received. If the spread between the (say) three lowest bids appears excessive or if the low bid is substantially below the architect-engineer's own estimate, this situation should be called to the attention of the low bidder. He should be asked to check his bid to determine whether there is an error in it, and if not, to verify its correctness in writing. If after verification the architect-engineer continues to feel that the proposal is too low, he should recommend that the owner exercise his right to reject the bid and accept another. The project can only suffer when it is let at too low a figure and the contractor has to resort to "mining the contract" (that is, searching for and pressing claims for extra work based on alleged errors and omissions in the Drawings and Specifications).

When a mistake in a proposal is discovered by a bidder after submission but before the time set for opening, ordinarily the bid may be withdrawn. When an error, however, is claimed after bids have been opened, the situation becomes sensitive. Unless the error is so gross that the owner could be charged with attempting to profit from a serious inadvertent mistake that both he and the architect should have detected (similar to the situation described in the preceding paragraph), the bidder may be obliged to accept the contract or forfeit his bid security.[5]

OTHER BID IRREGULARITIES.

Because of their frequent occurrence, state the instructions to bidders that the owner reserves the right to waive irregularities in the bids. Many of them are relatively inconsequential, and it would be pedantic and unrealistic to disqualify a bid because of them. Others, however, require careful consideration; among these are failure to furnish bid security, furnishing bid security in less than the required amount, and failure to acknowledge receipt of addenda.

The first is critical. If a bidder fails to furnish bid security in accordance with requirements of the instructions to bidders, his bid should be disqualified, unless the bond or check was forwarded separately from the proposal and its late arrival was due solely to a delay in the mails for which the bidder was not responsible.

[5]For a discussion of grounds on which a bid may be withdrawn, see Bernard Tomson, *It's the Law* (Great Neck, N Y: Channel Press, 1960), pp. 99-109.

When the amount of bid security furnished by the low bidder is less than that required in the instructions to bidders, this irregularity might be waived, providing that the amount furnished is equal to or greater than the difference between the low bid and the next acceptable bid.

Failure by the low bidder to acknowledge receipt of any particular addendum may or may not be a critical consideration. If the addendum in question did not affect the price, quality, or quantity of the work, the bidder may be permitted to supply the acknowledgment. However, if acknowledgment of an addendum that does affect the price, quality, or quantity of the work is missing, this irregularity should not be waived unless it is clear that the effect of the addendum on the price is trivial in comparison to the difference between the low bid and the next bid.

Other irregularities that occur from time to time and whose significance must be weighed are bids mailed but not received before the time of opening; bids not dated; bids not submitted with required number of copies; bids received without signature; and similar shortcomings either inadvertent or, in rare instances, possibly intended to create an equivocal situation.

Award of Contract

In private work where bids are received from a limited list of invited bidders the contract should be awarded to the lowest bidder if his bid is found to be in order. Although the owner has the right to reject all bids, this right should not be exercised as a subterfuge to permit an award to a contractor who did not submit a proposal before the results of the bidding were made public; nor should he use the bidding process as a mechanism for obtaining estimates and then negotiate with a "favorite" contractor who had been definitely selected in advance. This is an unfair practice and may result in responsible contracting firms refusing to bid on future work for that owner or on other projects of the architect-engineer. When an owner does a great deal of building, he has a stake in the good will of contractors and cannot expect them repeatedly to go through the expensive process of preparing bids if competition for the work does not in fact exist. Means of controlling this situation with "one-shot" clients, except through persuasion by architects and engineers, are, however, less apparent.

In public construction it is mandatory to make the award to the lowest "responsible" bidder, and the problem here is to establish and enforce practicable criteria. The matter of declaring a low bidder unacceptable is a very sensitive one and public awarding authorities are vulnerable to litigation initiated by disqualified low bidders unless a very strong case can be made to justify the action. It appears that default on a previous construction contract constitutes the strongest grounds for disqualification, but, of course, there are several other factors that are a measure of a bidder's capacity (or lack of it) to perform; but they may be difficult to sustain as grounds for rejection of his low bid on a particular project. This situation has led to requirements for prequalification in connection with bidding on much public work and on private work where circumstances are unfavorable to a limited list of invited bidders. Prequalification procedures are discussed in a subsequent article.

SUBCONTRACTOR APPROVAL

There is considerable difference of opinion about whether approval of subcontractors by the owner and architect should take place before or after award of the prime contract. There would seem to be little reason for requiring the listing of subcontractors in the proposal form unless it is intended to resolve the matter before the award. In any event, as noted earlier in the article on instructions to bidders, the procedure to be followed with respect to subcontractor approval should be specified in that document.

NOTICE OF AWARD

When the successful bidder has been determined, he should be sent an official notice of award. This notice constitutes acceptance by the owner of the bidder's proposal, and the way is then clear for executing the construction contract. Unless provision is made for issuing a formal *notice to proceed,* the contract time allowed for construction is considered to start as of the date of execution of the contract.

LETTERS OF INTENT

In construction work undertaken by large corporate owners and some public agencies, an appreciable delay often occurs between selection of the contractor and execution of the formal contract documents. In an effort to avoid delay in starting construction of an urgent project, the contractor may be given interim authorization to proceed under a *letter of intent.* Such a document is a limited order to proceed, and the latitude of action that the contractor has with respect to incurring costs and obligations to subcontractors may be severly circumscribed.

Explicit provisions must be incorporated to cover settlement costs in the event a formal contract is never executed and to assure effective insurance coverage while operations are progressing under the interim authorization. Letters of intent should be prepared for the owner's signature by the owner's attorney, with the participation of the architect-engineer limited to advice on technical matters for which he is responsible. When the formal contract is signed, it supersedes the letter of intent, and any payments that may have been made under the letter of intent are credited to the contract sum.

Prequalification of Contractors

An inherent premise of the competitive bidding system is that the finished building will be of substantially the same quality if constructed by any one of the bidders. This in turn implies that all bidders possess comparable capability and skill for performance of the work. Where bidding is limited to a carefully screened list of contractors known from experience to be of approximately equal competence and responsibility, these premises can be realized. In open bidding, however, whether on private or public construction, some mechanism is required for determining the "lowest responsible bidder."

In public work open bidding is usually mandatory, and in private work some owners insist on permitting contractors of differing reputation and skill to be included on the list (presumably hoping the weaker ones will not be low) because of business relationships not necessarily allied to the project. As noted earlier this pits the better builders against the poorer and may result in only courtesy bids from the better firms who feel a serious proposal is not worth the effort of preparation. Regardless of this point, however, the problems generated by not awarding the contract to the low bidder are generally far greater than those arising from not permitting him to bid in the first place. Prequalification procedures offer some measure of assistance in controlling the quality of bidders permitted to submit proposals on a particular project. This is why many states and several federal agencies have requirements concerning prequalification of contractors, and many architects and engineers urge the practice in private work where limited lists of selected bidders are not feasible for one reason or another.

Since the purpose to be served is the elimination of incompetent, overextended, and underfinanced contractors from the bidding, information about these factors is called for in advance. Statutes controlling public construction usually require that legal notice of prequalification requirements be published in designated newspapers and trade magazines in the same manner as advertisements for bids. In private work such public notice is not customary unless it is desired to solicit qualified bidders for a particular project.

PREQUALIFICATION PROCEDURE

This usually involves the execution of a questionnaire or qualification statement such as Form No. 40, promulgated by the Associated General Contractors of America, reproduced at the end of the chapter. Among other items of information called for by this questionnaire the contractor must reveal whether he has ever failed to complete any work awarded to him. As mentioned earlier, default on a previous contract presents about the strongest argument available for disqualifying a prospective bidder.

The listing of construction projects the contractor has under way as of the date of filing will indicate the scope of his current operations, and the value of major projects completed during the last few years is a measure of his average annual capacity. Comparison of these statements will indicate whether the contractor appears to have available capacity for new work without overextending his organization.

The information relating to construction experience of principal personnel may be indicative of the contractor's capability for coping with the type of project involved; and the statement of financial condition is intended to reveal the extent of his financial resources for carrying on the work.

One device for discouraging financially weak bidders is to require certified checks rather than bid bonds as bid security; another is to require certification from an acceptable bonding company that it will furnish performance and payment bonds if the bidder is awarded the contract. Obviously, inability to furnish these bonds constitutes adequate grounds for disqualification.

LIMITATIONS OF PREQUALIFICATION

The weakness of all prequalification procedures is the lack of any effective means of evaluating the more or less intangible factors of quality, technological know-how, and efficiency. These do not lend themselves satisfactorily to statistical comparison in the manner that financial rating, inventories of owned equipment, and size of organization do. This has led opponents of prequalification to charge that it promotes restriction of competition to the large and established contractors and tends to eliminate well-qualified younger firms. Others hold that in much public work the procedure degenerates to the level of routine, since the disqualification of a contractor is a serious action fraught with the risk of legal or at least political reprisal. Even though the basis of disqualification may be stated definitely as lack of specialized technical experience for the particular project, lack of adequate working capital, or insufficient operating capacity to undertake new work, there is an ever-present hazard that the action will be construed as a libelous reflection on the competency and integrity of the disqualified bidder. In spite of its shortcomings, however, prequalification is thought by many architects, engineers, and contractors to be a worthwhile procedure in the effort to eliminate bidders who are not qualified to carry out the work.

Bid Depositories

An essential phase of the bidding process is the solicitation and receipt of subcontractor bids by the prime bidders. Since as much as 80% of the contract price of a project constructed under the single contract system may represent work to be subcontracted, the owner and architect have more than a passing interest in this operation. Traditionally, within the building industry, business arrangements between contractor and subcontractor were of little concern to others than themselves. This provided an atmosphere conducive to developing some questionable practices in sub-bidding. Although the successful prime bidder might be assumed to have an obligation to award subcontracts to the firms whose bids he used in preparing his own, this procedure is not always followed. An unethical general contractor can attempt to lower the price of a sub-bid by playing off one sub-bidder against another. Of course, since it takes two parties to play this game (bid shopping), all blame cannot be attributed to the prime contractor even though he is in the stronger bargaining position. The listing of subcontractors on the proposal form represents one attempt to curtail practices of this nature and the establishment of bid depositories another.

Considerable variation in operating details exists among the several bid depositories (sometimes called bid registries) that have been established in the United States and Canada, but the general procedures are substantially similar. Where operative, the bids of subcontractors are sent to the depository in sealed envelopes with each one marked for the intended prime contractor. A deadline is established for submission of the sub-bids, which are then picked up by the prime bidders for comparison, analysis, and selection. A prime contractor is not necessarily obligated to use the lowest sub-bid submitted, but whichever one he uses he must accept at the price stated.

Bid depositories are usually operated by trade associations or builder's exchanges. Their merit and appropriateness are controversial in some quarters of the industry. Nevertheless, the persistence of unethical bidding procedures may well result in bid depositores of some nature playing an increasing role in bidding procedures.[6]

Separate Bidding

This procedure, denoting the taking of separate bids by the owner for mechanical and electrical work, is intended to achieve some of the purposes of the separate contracts system while retaining the centralized construction management characteristic of the single general contract. As noted in Chapter 1, a principal merit claimed for the separate contracts system is closer control by the owner over selection of firms to perform the electrical, plumbing, and heating and air conditioning work. A measure of control is provided on private work by listing in the instructions to bidders the names of subcontractors acceptable to the owner for these major divisions. The taking of separate bids, however, also disposes of the problem of bid shopping.

Although variations of the procedure exist, the intention is for the owner or awarding authority to call for four separate bids: the general construction, heating and air conditioning, plumbing, and the electrical work. After selection of subcontractors in the last three divisions, they are assigned to the successful bidder for the general construction, who then becomes the general contractor under a single contract with the owner. Various problems can arise when

[6]For an interesting discussion of bid depositores from different points of view see the article "Bid Shopping in Our Competitive Economy", *The Construction Specifier*, June 1966.

incompatibility exists among some of the bidders, since a general contractor cannot be expected to accept a subcontractor to whom he objects and vice versa. Furthermore, the general construction contractor must make allowance in his bid for his own costs in coordinating and managing the operations of the others, just as he would do if he obtained their sub-bids himself. Whether the separate bidding procedure will become a generally accepted alternative to the separate contracts system is not clear at this time.

Factors Influencing Level of Bids

In addition to the clarity and completeness of drawings and specifications from which a contractor's estimates are prepared, several other factors influence the level of bids received on a particular project. The most significant is the "bidding climate" or attitude of bidders during the bidding period. When a large volume of work is under contract and activity in architectural and engineering offices is great (with the prospect of many projects issuing for bid in the near future), bidders are provided leeway both in selecting and pricing jobs, and hence may not be in a competitive frame of mind. On the other hand, when much work under contract is nearing completion and new projects issuing for bid are in short supply, competition for the few jobs out for bid may be very keen. Whether this situation is brought about by general economic conditions or by seasonal variation in areas where weather is an important factor in construction activity, timing (either planned or fortuitous) may be responsible for a reduction in cost that is greater than any other economy measure.[7] Naturally, circumstances seldom permit taking full advantage of this situation, but it is one worth considering where tight budgets are involved and the nature of the project permits a degree of selective timing.

BIDDING PERIOD

The time allowed for preparation of bids can have a significant effect on their accuracy and therefore presumably on their level. When bidders are unduly pressed for time, they will be unable to obtain many competitive quotations from subcontractors and suppliers, much less to check them carefully. Faced with an unreasonable deadline, without full knowledge of their own costs, bidders must engage in a certain amount of guesswork. Prudence dictates that this be on the high side, with consequent inflation of the bid. As a rough guide, to be modified in accordance with the complexity of a particular project, a period of four weeks should be allowed for bidding on jobs estimated to fall within the range of one to two million dollars. It appears that failure to provide

[7] In this connection see *Potential Economies in School Building Construction*, W. F. Koppes, A. C. Green, and H. D. Haut, New York State Education Department, Albany, 1958.

sufficient time for preparation of bids may increase construction costs by as much as 5%.[8]

EXAMPLE OF CONTRACTOR'S QUALIFICATION FORM

Both the American Institute of Architects and the Associated General Contractors of America publish forms for use in prequalification procedures. The AIA Document A305—Contractor's Qualification Statement is similar to AGC Standard Form No. 40 presented here. Both documents are reciprocally approved and recommended by the two organizations.

Form No. 40 is reproduced by permission of AGC.

[8]See reference in footnote 7.

PRE-QUALIFICATION STATEMENT

(Required in advance of consideration of application to bid)

Approved and recommended by

AMERICAN INSTITUTE OF ARCHITECTS
AMERICAN SOCIETY OF MUNICIPAL ENGINEERS
THE ASSOCIATED GENERAL CONTRACTORS OF AMERICA, INC.

Submitted by _____

Address _____

Date _____

Note: If you have filed a Pre-Qualification Statement within the past_____
days, questions numbers 1, 14, and 15 only of this document need be an-
swered.

FIRST EDITION, 1933
(REPRINTED 1965 WITHOUT CHANGE)
ASSOCIATED GENERAL CONTRACTORS OF AMERICA
1957 E STREET, N.W., WASHINGTON, D.C. 20006

PRE-QUALIFICATION STATEMENT

Submitted to _____

By_____ $\left\{\begin{array}{l} \square \text{ A Corporation} \\ \square \text{ A Co-partnership} \\ \square \text{ An individual} \end{array}\right.$

Principal Office_____

The signatory of this questionnaire guarantees the truth and accuracy of all statements and of all answers to interrogatories hereinafter made

1. Have you filed a Pre-Qualification Statement within the past _____ days?_____
2. How many years has your organization been in business as a general contractor under your present business name? _____
3. How many years' experience in _____ construction work has your organization had: (a) As a General Contractor _____(b) as a Sub-Contractor_____ .
4. Corporation or Co-partnership Information:

If a corporation, answer this: Capital paid in cash, $ _____ When incorporated _____ In what State _____ President's name _____ Vice-President's name _____ Secretary's name _____ Treasurer's name _____	If a co-partnership, answer this: Date of organization_____ State whether partnership is general, limited or association _____	
	Name and address of partners:	Age
	_____	___
	_____	___

5. List the construction projects your organization has under way on this date:

Contract Amt.	Class of work	Per cent Completed	Name and address of Owner or Contracting Officer

6. List projects your organization has completed in past three years:

Contract Amt.	Class of work	When Completed	Name and address of Owner

Use blank sheet if additional space is needed.

7. Have you ever failed to complete any work awarded to you? _____If so, where
and why?_____

8. Has any officer or partner of your organization ever been an officer or partner of some
   other organization that failed to complete a construction contract? _____
   If so, state name of individual, other organization and reason therefor _____
   _____
   _____

9. Has any officer or partner of your organization ever failed to complete a construction
   contract handled in his own name? _____If so, state name of individual, name of
   Owner and reason therefor _____
   _____
   _____

10. In what other lines of business are you financially interested? _____
    _____
    _____

11. What is the construction experience of the principal individuals of your organization?

Individual's Name	Present Position or Office	Years of Construction Experience	Magnitude and Type of Work	In What Capacity

12. What equipment do you own that is available for proposed work?

Quantity	Item	Description, Size, Capacity, Etc.	Condition	Years of Service	Present Location

13. Give condensed current financial statement:

Condition at close of business ———————— 19 ————					
ASSETS	Dollars			Cts.	
1. Cash: (a) On hand $ ——— , (b) In bank $ ——— , (c) Elsewhere $ ——					
2. Notes receivable (a) Due within 90 days ———————					
(b) Due after 90 days ———————					
(c) Past due ———————					
3. Accounts receivable from completed contracts, exclusive of claims not approved for payment ———————					
4. Sums earned on uncompleted contracts as shown by Engineer's or Architect's estimate					
(a) Amount receivable after deducting retainage———					
(b) Retainage to date, due upon completion of contracts———					
5. Accounts receivable from sources other than construction contracts ———————					
6. Deposits for bids or other guarantees:					
(a) Recoverable within 90 days ———————					
(b) Recoverable after 90 days ———————					
7. Interest accrued on loans, securities, etc.					
8. Real estate: (a) Used for business purposes ———					
(b) Not used for business purposes———					
9. Stocks and bonds: (a) Listed —present market value ———					
(b) Unlisted — present value ———					
10. Materials in stock not included in Item 4					
(a) For uncompleted contracts (pres. value) ———					
(b) Other materials (present value) ———					
11. Equipment, book value———————					
12. Furniture and fixtures, book value ———————					
13. Other assets ———————					
Total assets———					
LIABILITIES					
1. Notes payable: (a) To banks regular———————					
(b) To banks for certified checks———					
(c) To others for equipment obligations ———					
(d) To others exclusive of equipment obligations ———					
2.*Accounts Payable: (a) Not past due ———————					
(b) Past due ———————					
3. Real estate encumbrances ———————					
4. Other liabilities ———————					
5. Reserves ———————					
6. Capital stock paid up: (a) Common ———————					
(b) Common ———					
(c) Preferred ———					
(d) Preferred ———					
7. Surplus (net worth) Earned $ ———Unearned $ ———					
Total liabilities					

13. (Cont.)

CONTINGENT LIABILITIES	Dollars				Cts.
1. Liability on notes receivable, discounted or sold ————					
2. Liability on accounts receivable, pledged, assigned or sold——					
3. Liability as bondsman ————————————					
4. Liability as guarantor on contracts or on accounts of others —					
5. Other contingent liabilities ————————————					
Total contingent liabilities					

*Include all amounts owing subcontractors for all work in place and accepted on
completed and uncompleted contracts, including retainage.

14. Will you, upon request, fill out an approved form of detailed financial statement and an
additional form of Job Plan and Equipment Questionnaire? ————————————

15. Have you filed Performance Record reports with the Bureau of Contract Information,
Inc., Washington, D. C.? ————————————————————————————

Dated at ————————————————————————this ——————————
day of ———————————— 19——

————————————————————
Name of Organization

By ————————————————————

Title of Person Signing

STATE OF ————————————— }
                                    } ss.:
COUNTY OF————————————— }

——————————————————— being duly sworn deposes and says that he

is ———————— of————————————————————————

Name of Organization

and that the answers to the foregoing questions and all statements therein contained are
true and correct.

Sworn to before me this

———————————— day of ————19——

————————————————————
Notary Public

My commission expires ————————————

Chapter 8

Lump-Sum Construction Agreements and Subcontracts

Nature of the Agreement

The Agreement is the basic instrument that welds all the Construction Documents into a single integrated contract. It is a relatively brief document whose principal functions are to identify specifically all other elements constituting the contract, to state the contract price and contract time, and to serve as the agent for its formal execution. Although it is common to include additional provisions in the Agreement, many of them are of such nature that they could just as well appear in the Supplementary Conditions. Consequently, practice varies with respect to content and arrangement, but articles covering the following items will usually be found in the Agreement:

Identification of Parties and Project
Identification of Contract Documents
Contract Time
Liquidated Damages (when applicable)
Contract Sum
Progress Payments
Final Payment
Execution of Agreement

As in earlier chapters this discussion is independent of any standard form, but two such forms are presented at the end of the chapter.[1] It will be of interest to

[1] These are AIA Document A101, "Standard Form of Agreement Between Owner and Contractor where the basis of payment is a Stipulated Sum" and AIA Document A107, "Short Form Agreement for Small Construction Contracts."

the reader to compare the provisions in these standard documents with the discussion of corresponding topics that follow. Again, no direct reference to article numbers is made, but a brief examination will show the arrangement of articles and facilitate cross reference. Lump-Sum Agreements will be considered first with reference to construction carried on under a single general contract.

Identification of Parties and Project

This information usually appears as a preamble to the formal articles and includes the date "as of" that the Agreement is executed. It should be made clear to both owner and contractor whether each is doing business as an individual, partnership, joint venture, or corporation, and the legal addresses of both parties should be given. Although the architect is not a party to the construction contract, he performs several functions thereunder as provided in the General Conditions and should be identified by name in the Agreement.

Identification of the project is usually covered adequately by citing the descriptive title used on the Drawings and Specifications (for example, Blank Airlines Maintenance Facility, Los Angeles International Airport, Los Angeles, California). More precise information concerning location can be given on the site plan or elsewhere. Of course, if the project is in a remote location, even the general descriptive title might include the county or township for easy recognition.

Identification of Contract Documents

It is customary to accomplish this identification in the first numbered article of the Agreement. In addition to stipulating that the Contract Documents consist of "this" Agreement, the General Conditions, Supplementary Conditions, Drawings, Specifications, and Addenda issued before execution, provision is also made to assure that all Modifications (change orders, supplemental amendments, etc.) that may be made subsequent to execution of the Agreement will become a part thereof. After citing the Contract Documents in this manner, a statement is made to the effect that ". . . these form the contract, and are all as fully a part of the contract as if attached to this Agreement or repeated herein".

It is necessary that the Agreement contain a more detailed enumeration of the Contract Documents than the foregoing listing or it should be provided in the Supplementary Conditions and cited specifically in the Agreement. Each document for which a standard form is not employed should be listed in order to indicate the number of pages involved. This will ordinarily be required for the Supplementary Conditions, any additional special conditions that may be included, and for each division of the Specifications. Also the sheet numbers of the various sets of drawings (architectural, structural, mechanical, electrical, etc.)

must be given. When Addenda are involved, the number and date of each Addendum should be recorded, and if any alternates were accepted under the notice of award, each should be identified by number and the pertinent "alternative specifications and drawings" cited.

Contract Time

As noted under discussion of award procedures in Chapter 7, the contract time is considered to start as of the date of execution of the Agreement unless provision is made otherwise. Frequently, the bid form is so drawn that the successful bidder is pledged to commence work "immediately on execution of the contract." When time of performance is to be strictly enforced and when a substantial delay between opening of bids and signing the contract may be expected, this provision may prove unfair to the contractor. He must either maintain at least partial mobilization of his resources over an uncertain length of time in order to be able to start immediately or risk cutting into the contract time for preparatory activity after the contract is executed. It is for this reason that the pledge to start work is more equitable when there is provision for the contractor to commence work within a stipulated number of days (say five or ten) after execution of the contract. This can be of considerable importance when liquidated damages provisions, as discussed later, are part of the contract.

This matter can be handled with more precision by the device of a *notice to proceed*, prepared by the owner and dated as appropriate. Instead of stipulating a fixed date on which the work is to start, the pertinent article of the Agreement would then read, ". . . shall commence on the date stipulated in the notice to proceed and shall be substantially completed on . . . ." This effectively establishes the starting date as well as the contract time allowed for completion. The notice to proceed should, of course, never be issued before execution of the contract. If the urgency is such that the owner desires to begin operations before contract signing, he should make use of a letter of intent.

It is neither necessary nor appropriate to make any mention in the Agreement relating to procedures for granting possible extensions of time; these matters are covered in detail in the General Conditions, which also define substantial completion for purposes pertaining to contract time. It is widely recognized that time limits are difficult to enforce. However, if they are not included in the Agreement, there may be little basis for initiating termination proceedings in event of unreasonable delay in the contractor's prosecution of the work.

Liquidated Damages

Where time is of the essence of a contract and the project is of such nature that the owner will incur loss of revenue or other provable loss if completion is delayed beyond the time limits stated in the Agreement, the owner may be in a legal position to recover damages from the contractor. In order to avoid the necessity for the owner to bring a lawsuit against the contractor to prove such loss, liquidated damages may be stipulated in the Agreement. These consist of a stated dollar amount per day of delay that the contractor will pay the owner in lieu of determining the actual amount of damages suffered. This amount must be based on a reasonable estimated loss to the owner if the project is not finished on time (for example, in a department store, loss of estimated net sales per day of delay in opening). When the amount of liquidated damages is so determined and stipulated in the Agreement, it is enforceable at law. Excessive liquidated damages implies a penalty, and a penalty will not be assessed unless there is a corresponding bonus provision for early completion, as discussed later.

Where liquidated damages are to be stipulated in the Agreement, a provision similar to the following may be used:

> "For each calendar day in excess of the established completion date that the Work remains incompleted the contractor agrees to pay to the owner, the sum of $ _____ as liquidated damages, which are reasonably estimated in advance to cover the losses to be incurred by the owner by reason of failure of the contractor to complete the Work, time being of the essence of the contract and a material consideration thereof."[2]

It should be noted that the decision to assess liquidated damages may have an effect on the level of bids received. If a bidder feels that the time allowed for construction is unrealistically short, he must either add an allowance to cover anticipated liquidated damages or in his estimate provide for the additional costs involved for overtime or multiple-shift operations.

PENALTY AND BONUS

To supply an incentive for the contractor to meet or anticipate the scheduled completion date, provision may be made for paying a bonus for each day of earlier completion. An accompanying provision would require a similar amount to be paid by the contractor as a penalty for each day of delay beyond scheduled completion. Here again the dollar amount established must be a reasonable measure of benefits that would accrue to the owner by virtue of early completion or be lost if completion were delayed.

Where a penalty and bonus clause is to be used in lieu of a liquidated damages provision, it may read as follows:

[2]Architect's Handbook of Professional Practice, AIA, Chapter 17 (1966 Interim Printing), p 10.

"The contractor agrees to pay the owner a sum of $ _____ for each day beyond the estimated completion date that the Work remains incompleted, in consideration of which the owner agrees to pay the contractor a sum of $ _____ for each day ahead of the established completion date that the Work is completed."[3]

The merits of penalty and bonus provisions in construction contracts are controversial. Although they are frequently encountered, some architects and engineers feel they tend to encourage speculation and may turn out to work to the disadvantage of both owner and contractor.

Contract Sum

The principal items to be covered under this article are the amount of the contract sum at the time of executing the contract, recognition of the fact that it is subject to adjustment, and listing of agreed-upon unit prices if any. In the first instance the contract sum would be the amount of the bid as modified by acceptance or rejection of alternates and by negotiation of minor changes between the time of receiving bids and making the award. This is stated as a lump-sum amount.

Recognition that the contract sum is subject to adjustment is usually handled by a phrase such as ". . . subject to additions and deductions by change order as provided in the Conditions of the Contract. . ." As discussed in Chapter 4, changes may be initiated by the owner as he sees fit and by the contractor in connection with claims for extra work arising from interpretation of the Contract Documents or from other sources. Thus there may be several adjustments in the contract sum as the work progresses.

Unit prices to appear under this article are those that may have been called for in the bids or subsequently negotiated to facilitate determination of costs of changes that may be required in certain elements of the project. In addition to the listing of these unit prices, a clause should be incorporated stipulating the conditions under which they are to be applied.

Progress Payments

The mechanics of making progress payments are stated in the General Conditions, but in order to implement them certain determinations must be made and stated in the Agreement: due dates for payments, amount of retained percentage for both work in place and stored materials and equipment, the cutoff date, and special considerations relating to the payment made at time of substantial

[3]See reference in footnote 2.

completion. With respect to the first three factors, the Agreement provides that the owner will make progress payments on a certain day of each month in an amount equal to a stated percentage of the value of the work accomplished up to a cutoff date occurring a specified number of days earlier, less, of course, the amount of payments made previously.

The due date established should be mutually acceptable to contractor and owner and should consider the time required for preparation of an application for payment, its checking and certification by the architect, and for the owner to make payment within the time limits set in the General Conditions.

RETAINED PERCENTAGE

The amount of retained percentage varies, but 10% is a widely used rate for both work in place and stored materials and equipment. This, of course, results in each progress payment being made as 90% of the amount approved on the contractor's application. Unless other provision is made, this same percentage is applied to all progress payments until substantial completion which, for large projects, amounts to a considerable sum.

When projects are progressing well and 50% completion has been achieved, it may appear that 10% retainage on the first half of the work (representing 5% of the contract sum) is adequate. If both owner and surety concur, no additional retainage is deducted or, in some instances, a reduced percentage is applied to the last half of the job. Any provisions for limiting or reducing the amount retained after the work reaches a certain stage should appear in the progress payments article of the Agreement, and should be so worded that their implementation depends on the judgment of the architect-engineer and written consent of the surety.

PAYMENT AT SUBSTANTIAL COMPLETION

The procedure leading to certification of substantial completion of the project is prescribed in the General Conditions, but the Agreement should include a statement of the percentage of the contract sum due at that time. This frequently takes the form "... and upon substantial completion of the work, a sum sufficient to increase the total payments to (say 95) percent of the contract sum, less such retainages as the architect-engineer shall determine for incomplete work and unsettled claims." It should be noted that the total amount retained at the time of substantial completion may be greater than the (in this case) implied 5% due to known outstanding commitments that do not interfere with beneficial use of the building under the definition of substantial completion given in the General Conditions.

Final Payment

This is usually a short article since all the specific conditions that must be fulfilled before a final certificate for payment is issued are stated in detail in the

General Conditions, as discussed in Chapter 4. The essential function of the statement in the Agreement is to establish a due date for final payment with reference to acceptance of the project by the owner. This may be accomplished by stipulating that the final payment which constitutes the unpaid balance of the contract sum ". . . shall be paid (say 30) days after substantial completion unless otherwise provided in the certificate of substantial completion, and provided the work has then been completed and a final certificate for payment has been issued. . . ."

The widely used thirty-day period between substantial completion and the date final payment is due is intended to provide adequate time to accomplish all work on the "punch list" and to assemble all releases, certifications, and other documents required by the Conditions of the Contract as prerequisites to final payment. It is necessary to stipulate some time limit in order to protect the contractor against unreasonable delay in receiving his final payment. On the other hand, as a protection for the owner, the obligation to make final payment within the stipulated period is safeguarded by the requirement that the work must be completed by the end of the period.

It will be recalled from the discussion in Chapter 4 that provision is normally included in the General Conditions to cover cases where final completion is materially delayed through no fault of the contractor. Implementation of such provisions permits payment to be made for that portion of the work that is completed and accepted without terminating the contract.

Execution of Agreement

The two items of principal concern here are the date of the Agreement and the validity of the authority of those signing it. Since the form is usually prepared somewhat ahead of the time of actual execution, there is a possibility that the date on the face of the Agreement may not be the same as the date of signing. It will be noted that both contract forms reproduced at the end of the chapter avoid this possible ambiguity by citing the date first mentioned in the Agreement as the effective date of execution. This date should not, of course, be later than the date on which work actually begins. If work is started under a letter of intent ahead of the time a formal contract is signed, the date on the Agreement when subsequently executed should be that of the letter of intent. This is necessary since, as noted earlier, all obligations stipulated in the letter are taken over by the formal contract.

Practice varies with respect to the form in which signatures are to appear in the Agreement and should be determined by the owner and his attorney. Some contracting agencies require addresses in the signature blocks, and provision for signatures of witnesses is also made in some instances. However, absence of witnesses' signatures does not affect the validity of the contract. When a company name is used, it is followed by the signature of an individual and his title, that is, individual proprietor, member of a partnership, or officer of a corporation. If the officer of a corporation signs, his authority should be attested by signature of the secretary or other appropriate officer. Special consideration is required with contracts signed on behalf of public bodies in order to assure that requisite action authorizing the contract has been taken and recorded.

Separate Contracts Agreements

When a project is to be built under the separate contracts system, a separate set of Contract Documents must be prepared for each contract. This presents no special problem with regard to Drawings and Specifications, since Drawings are frequently broken into sets for convenience and Specifications often have the mechanical and electrical divisions presented in separate volumes. Carrying this procedure a step further readily yields separate Drawings and Specifications for each of the four separate contracts customarily involved: general construction, heating and air conditioning, plumbing, and electrical.

It is advantageous to have identical General Conditions for all the multiple prime contracts, with necessary additions adapting them to the special requirements of each contract presented in separate Supplementary Conditions. This can theoretically be achieved with some of the standardized general conditions now published, but many architects and engineers feel that the substantial modifications and amendments required to do so results in overly complex Conditions of the Contract that are difficult to understand. Where separate General Conditions are to be prepared, their provisions should be made as nearly alike as possible and special attention given to provision for coordination of the work of the separate contractors. This all-important factor under the separate contracts system should also be made the subject of a special article in each separate Agreement. Other articles of the separate Agreements cover subject matter similar to those previously identified and will be considered below.

IDENTIFICATIONS

Information pertaining to identification of parties and project is similar to that required under the single contract system, each Agreement in this case using the identical descriptive title (for example, Blank Airlines Maintenance Facility, Los Angeles International Airport, Los Angeles, California). The name of the owner will be stated identically in the preambles of all Agreements, and the separate contractor involved will be identified by his firm's official name and address in his Agreement only. In order to facilitate reference in the several Agreements to the *other* prime contractors, official designations may be established such as Plumbing Contractor, General Construction Contractor, and Electrical

Contractor without additional identification. When the designation "the Contractor" is used, this is defined as the contractor who signs the Agreement in question.

With respect to identification of the Contract Documents, this must be sufficiently definite to differentiate them from those of the other separate prime contracts. It can be achieved by being very specific in the enumeration that supplements the general listing, as shown below for the general construction separate contract:

This Agreement
General Conditions of the Contract for General Construction
Supplementary Conditions of the Contract for General Construction
Drawings as follows:

Architectural	Nos. A1 through A30
Structural	Nos. S1 through S16

Specifications as follows:

General Requirements	Division 1	13 pages
Site Work	Division 2	5 pages
Concrete	Division 3	25 pages
etc. . .		

Particular attention must be paid to the listing of addenda under the separate contracts system since any one addendum issued may or may not affect all prime contracts.

CONTRACT TIME

An additional factor that must be covered in this article is provision for some commitment on the part of the separate prime contractor to collaborate with the overall construction schedule (whoever may have the responsibility for establishing it) and time his own operations accordingly. Technically the problem is no more difficult than that faced by a single general contractor in scheduling the work of subcontractors, but administrative complexities can arise because no direct contractual relationship exists among any of the separate contractors. The independent responsibilities of the separate contractors also makes assessment of liquidated damages a difficult procedure.

CONTRACT SUM AND PAYMENTS

Articles covering these subjects can be substantially the same as those in an Agreement relating to a single contract, although the wording may require modification if the path for processing applications for payment is other than directly from the separate contractors to the architect. This might be the case where special arrangements are established for overall management and coordination of the separate prime contractors.

Short Form Contracts

For many small projects an abridged form of the Contract Documents may be appropriate. In recognition of this situation, the American Institute of Architects has issued a short form, lump-sum contract as AIA Document A107. The full title of this document, reproduced at the end of the chapter, is "Short Form Contract for Small Construction Contracts Where the Basis of Payment is a Stipulated Sum." It combines in one document an Agreement and selected provisions from the AIA standardized General Conditions which, together with the Drawings and Specifications, constitute the entire contract.

Short form contracts should be used with caution since many of the situations covered by articles omitted in the abridgement may, of course, arise on small projects as well as on large ones. The limitations of the short form will be revealed by a comparative study of AIA Documents A107 and A201.

Subcontracts

Although both owner and architect have an obvious interest in the quality of subcontractors chosen for the project, neither has any contractual relationship with the subcontractors. As noted in Chapter 4, the principal reason for treating this subject at all in the General Conditions is to provide the owner with some measure of control over subcontracting operations. This is achieved indirectly by stipulating in the prime contract certain procedures the contractor is to follow in executing and administering his contracts with the subcontractors. Because the scope of work to be accomplished under different subcontracts varies widely, informal letters of proposal and acceptance are frequently utilized in lieu of more formal documents. When this procedure is followed, however, judgment must be exercised by the contractor, since he remains obligated to bind his subcontractors to himself in accordance with the terms of the General Conditions of the prime contract.

Formal subcontracts usually combine in one document the covenants of an agreement and the general terms and conditions of the subcontract. The principal items that require attention are

Recognition of prime contract provisions
Identification of subcontractor's work
Subcontract time and schedule
Payments to the subcontractor
Special subcontract provisions
General terms and conditions
Execution of subcontract agreement

A brief comment on each follows, but it is suggested that the reader identify related clauses in the illustrative guide subcontract form reproduced at the end of the chapter.

Recognition of Prime Contract Provisions

This usually takes the form of a statement to the effect that the subcontractor agrees to perform all of the work (identified as part of the subcontract) in accordance with the Contract Documents constituting the contract between owner and prime contractor. This is customarily established further by making the General Conditions, Supplementary Conditions, Drawings, Specifications, and Addenda of the prime contract part of the subcontract by reference.

Identification of Subcontractor's Work

A precise description of the work appears under this article and it should be carried out as far as possible by reference to applicable sheet numbers of the Drawings and page numbers of the Specification. It is well to recall here that dividing the prime contract into subcontracts is a responsibility of the prime contractor. Although careful attention to the content of specification sections when being prepared by the architect-engineer will be helpful, the responsibility for establishing the scope of subcontracts remains that of the prime contractor.

Subcontract Time and Schedule

The starting date for a subcontract will, of course, depend on the nature of the work involved. In order to meet his own time commitments and to coordinate operations at the site, the prime contractor must establish starting dates for all subcontracts in accordance with requirements of the project construction schedule. When these dates are not determined before the execution of a particular subcontract, the procedure to be used in notifying the subcontractor to start work must be described. It is also customary to include a clause whereby the subcontractor agrees to prosecute his work in accordance with the overall progress schedule prepared by the prime contractor. This article of the subcontract is also the place to stipulate any liquidated damages provisions that may be applicable.

Payments to the Subcontractor

By terms of the General Conditions of the prime contract the subcontractor is to be paid at the same time that the owner makes periodic payments to the prime contractor. This article should stipulate the day of the month when the subcontractor must file his application for payment, any provisions for including cost of materials and equipment purchased and stored but not yet in place, retainage, etc. It should be noted that most subcontracts do not contemplate payment by the prime contractor ahead of the time he is paid by the owner, but this does not necessarily imply that the subcontractor's reimbursement is contingent on the owner paying the prime contractor. Final payment to the subcontractor may be made conditional on his furnishing evidence that all known indebtedness connected with his work on the project has been satisfied. This parallels the prime contractor's similar commitment to the owner before a certificate of final completion is issued.

Special Subcontract Provisions

The General Conditions of the prime construction contract frequently require that certain provisions of the prime contract be reproduced in all subcontracts. This applies particularly to federal and federally assisted projects and to much other public works construction. Among these provisions are those requiring conformance with prevailing wage rate determinations (David-Bacon Act, the Eight-Hour Laws, and Equal Employment Opportunity Clauses).

Special articles of the subcontract are sometimes devoted to cooperative arrangements concerning labor matters for the project, use of temporary site facilities (hoists, scaffolding, sheds, heat, water, etc.), and other matters contingent on the character and location of a particular project. Most other operating provisions of the subcontract, however, are placed in the general terms and conditions.

General Terms and Conditions

These constitute a sort of "general conditions" of the subcontract. They may be more or less extensive depending on the particular trade specialty involved and the degree to which the General Conditions of the prime contract specify procedures the prime contractor must follow in handling his subcontracts. In general, items treated under this heading represent those that experience has demonstrated must be covered to avoid disputes on the job.

Execution of Subcontracts

The same considerations apply here as those pertaining to the execution of any Agreement as discussed earlier in the chapter. However, if terms of the prime contract require that the owner approve the subcontractors, the prime contractor must exercise care to see that such approval is obtained before execution of any individual subcontract.

EXAMPLES OF STANDARD FORMS

Several standard forms pertaining to lump-sum agreements and subcontracts are presented here for illustrative purposes. Reproduction of copyrighted forms is by permission, respectively, of the American Institute of Architects and the Associated General Contractors of America, Inc.

THE AMERICAN INSTITUTE OF ARCHITECTS

AIA Document A101

Standard Form of Agreement Between Owner and Contractor

where the basis of payment is a

STIPULATED SUM

Use only with the latest Edition of AIA Document A201, General Conditions of the Contract for Construction.

AGREEMENT

made this   day of   in the year of Nineteen
Hundred and

BETWEEN

the Owner, and

the Contractor.

The Owner and the Contractor agree as set forth below.

AIA DOCUMENT A101 • OWNER-CONTRACTOR AGREEMENT • SEPTEMBER 1967 EDITION
AIA® © THE AMERICAN INSTITUTE OF ARCHITECTS, 1735 NEW YORK AVENUE, N.W., WASH., D.C. 20006   1

ARTICLE 1

THE CONTRACT DOCUMENTS

The Contract Documents consist of this Agreement, Conditions of the Contract (General, Supplementary and other Conditions), Drawings, Specifications, all Addenda issued prior to execution of this Agreement and all Modifications issued subsequent thereto. These form the Contract, and all are as fully a part of the Contract as if attached to this Agreement or repeated herein. An enumeration of the Contract Documents appears in Article 8.

ARTICLE 2

THE WORK

The Contractor shall perform all the Work required by the Contract Documents for
(Here insert the caption descriptive of the Work as used on other Contract Documents.)

ARTICLE 3

ARCHITECT

The Architect for this Project is

ARTICLE 4

TIME OF COMMENCEMENT AND COMPLETION

The Work to be performed under this Contract shall be commenced

and completed
(Here insert any special provisions for liquidated damages relating to failure to complete on time.)

ARTICLE 5

CONTRACT SUM

The Owner shall pay the Contractor for the performance of the Work, subject to additions and deductions by Change Order as provided in the Conditions of the Contract, in current funds, the Contract Sum of

(State here the lump sum amount, unit prices, or both, as desired.)

ARTICLE 6

PROGRESS PAYMENTS

Based upon Applications for Payment submitted to the Architect by the Contractor and Certificates for Payment issued by the Architect, the Owner shall make progress payments on account of the Contract Sum to the Contractor as provided in the Conditions of the Contraxt as follows:

On or about the   day of each month   per cent of the proportion of the Contract Sum properly allocable to labor, materials and equipment incorporated in the Work and   per cent of the portion of the Contract Sum properly allocable to materials and equipment suitably stored at the site or at some other location agreed upon in writing by the parties, up to the   day of that month, less the aggregate of previous payments in each case; and upon Substantial Completion of the entire Work, a sum sufficient to increase the total payments to   per cent of the Contract Sum, less such retainages as the Architect shall determine for all incomplete Work and unsettled claims.
(Here insert any provisions made for limiting or reducing the amount retained after the Work reaches a certain stage of completion.)

ARTICLE 7

FINAL PAYMENT

Final payment, constituting the entire unpaid balance of the Contract Sum, shall be paid by the Owner to the Contractor   days after Substantial Completion of the Work unless otherwise stipulated in the Certificate of Substantial Completion, provided the Work has then been completed, the Contract fully performed, and a final Certificate for Payment has been issued by the Architect.

ARTICLE 8

MISCELLANEOUS PROVISIONS

8.1   Terms used in this Agreement which are defined in the Conditions of the Contract shall have the meanings designated in those Conditions.

8.2   The Contract Documents, which constitute the entire agreement between the Owner and the Contractor, are listed in Article 1 and, except for Modifications issued after execution of this Agreement, are enumerated as follows:
(List below the Agreement, Conditions of the Contract (General, Supplementary, other Conditions), Drawings, Specifications, Addenda and accepted Alternates, showing page or sheet numbers in all cases and dates where applicable.)

This Agreement executed the day and year first written above.

OWNER_____ CONTRACTOR _____

THE AMERICAN INSTITUTE OF ARCHITECTS

AIA Document A107

The Standard Form of Agreement Between Owner and Contractor

Short Form Agreement for **Small Construction Contracts**

Where the Basis of Payment is a

STIPULATED SUM

For other contracts the AIA issues Standard Forms of Owner-Contractor Agreements and Standard General Conditions of the Contract for Construction for use in connection therewith.

AGREEMENT

made this   day of   in the year Nineteen
Hundred and

BETWEEN

the Owner, and

the Contractor.

The Owner and Contractor agree as set forth below.

AIA DOCUMENT A107 · SMALL CONSTRUCTION CONTRACT · SEPTEMBER 1966 EDITION · AIA®
©THE AMERICAN INSTITUTE OF ARCHITECTS, 1735 N.Y. AVE., N.W., WASHINGTON, D.C. 20006   1

ARTICLE 1

THE WORK

The Contractor shall perform all the Work required by the Contract Documents for
(Here insert the caption descriptive of the Work as used on other Contract Documents.)

ARTICLE 2

ARCHITECT

The Architect for this project is

ARTICLE 3

TIME OF COMMENCEMENT AND COMPLETION

The Work to be performed under this contract shall be commenced

and completed

ARTICLE 4

CONTRACT SUM

The Owner shall pay the Contractor for the performance of the Work, subject to additions
and deductions by Change Order as provided in the General Conditions, in current funds,
the Contract Sum of
(State here the lump sum amount, unit prices, or both, as desired.)

ARTICLE 5

PROGRESS PAYMENTS

Based upon Applications for Payment submitted to the Architect by the Contractor and
Certificates for Payment issued by the Architect, the Owner shall make progress payments
on account of the Contract Sum to the Contractor as follows:

ARTICLE 6

FINAL PAYMENT

The Owner shall make final payment   days after completion of the Work, provided
the Contract be then fully performed, subject to the provisions of Article 17 of the General
Conditions.

ARTICLE 7

ENUMERATION OF CONTRACT DOCUMENTS

The Contract Documents are as noted in Paragraph 8.1 of the General Conditions and are
enumerated as follows:
(List below the Agreement, Conditions of the Contract (General, Supplementary, and other
Conditions), Drawings, Specifications, Addenda and accepted Alternates, showing page or
sheet numbers in all cases and dates where applicable.)

GENERAL CONDITIONS

ARTICLE 8

CONTRACT DOCUMENTS

8.1   The Contract Documents consist of this Agreement (which includes the General Conditions), Supplementary and other Conditions, the Drawings, the Specifications, all Addenda issues prior to the execution of this Agreement, all amendments, Change Orders, and
written interpretations of the Contract Documents issued by the Architect. These form the
Contract and what is required by any one shall be as binding as if required by all. The
intention of the Contract Documents is to include all labor, materials, equipment and other
items as provided in Paragraph 11.2 necessary for the proper execution and completion of
the Work and the terms and conditions of payment therefor, and also to include all work
which may be reasonably inferable from the Contract Documents as being necessary to
produce the intended results.

8.2   The Contract Documents shall be signed in not less than triplicate by the Owner and
the Contractor. If either the Owner or the Contractor do not sign the Drawings, Specifications, or any of the other Contract Documents, the Architect shall identify them. By
executing the Contract, the Contractor represents that he has visited the site and familiarized himself with the local conditions under which the Work is to be performed.

8.3   The term Work as used in the Contract Documents includes all labor necessary to
produce the construction required by the Contract Documents, and all materials and equipment incorporated or to be incorporated in such construction.

ARTICLE 9

ARCHITECT

9.1   The Architect will provide general administration of the Contract and will be the Owner's representative during the construction period.

9.2   The Architect shall at all times have access to the Work wherever it is in preparation and progress.

9.3   The Architect will make periodic visits to the site to familiarize himself generally with the progress and quality of the Work and to determine in general if the Work is proceeding in accordance with the Contract Documents. On the basis of his on-site observations as an Architect, he will keep the Owner informed of the progress of the Work, and will endeavor to guard the Owner against defects and deficiencies in the Work of the Contractor. The Architect will not be required to make exhaustive or continuous on-site inspections to check the quality or quantity of the Work. The Architect will not be responsible for construction means, methods, techniques, sequences or procedures, or for safety precautions and programs in connection with the Work, and he will not be responsible for the Contractor's failure to carry out the Work in accordance with the Contract Documents.

9.4   Based on such observations and the Contractor's Applications for Payment, the Architect will determine the amounts owing to the Contractor and will issue Certificates for Payment in accordance with Article 17.

9.5   The Architect will be, in the first instance, the interpreter of the requirements of the Contract Documents. He will make decisions on all claims and disputes between the Owner and the Contractor. All his decisions are subject to arbitration.

9.6   The Architect has authority to reject Work which does not conform to the Contract Documents and to stop the Work, or any portion thereof, if necessary to insure its proper execution.

ARTICLE 10

OWNER

10.1   The Owner shall furnish all surveys.

10.2   The Owner shall secure and pay for easements for permanent structures or permanent changes in existing facilities.

10.3   The Owner shall issue all instructions to the Contractor through the Architect.

ARTICLE 11

CONTRACTOR

11.1   The Contractor shall supervise and direct the Work, using his best skill and attention. The Contractor shall be solely responsible for all construction means, methods, techniques, sequences and procedures and for coordinating all portions of the Work under the Contract.

11.2   Unless otherwise specifically noted, the Contractor shall provide and pay for all labor, materials, equipment, tools, construction equipment and machinery, water, heat, utilities, transportation, and other facilities and services necessary for the proper execution and completion of the Work.

11.3   The Contractor shall at all times enforce strict discipline and good order among his employees, and shall not employ on the Work any unfit person or anyone not skilled in the task assigned to him.

11.4   The Contractor warrants to the Owner and the Architect that all materials and equipment incorporated in the Work will be new unless otherwise specified, and that all Work will be of good quality, free from faults and defects and in conformance with the Contract Documents. All work not so conforming to these standards may be considered defective.

11.5   The Contractor shall pay all sales, consumer, use and other similar taxes required by law and shall secure all permits, fees and licenses necessary for the execution of the Work.

11.6   The Contractor shall give all notices and comply with all laws, ordinances, rules, regulations, and orders of any public authority bearing on the performance of the Work, and shall notify the Architect if the Drawings and Specifications are at variance therewith.

11.7   The Contractor shall be responsible for the acts and omissions of all his employees and all Subcontractors, their agents and employees and all other persons performing any of the Work under a contract with the Contractor.

11.8   The Contractor shall furnish all samples and shop drawings as directed for approval of the Architect for conformance with the design concept and with the information given in the Contract Documents. The Work shall be in accordance with approved samples and shop drawings.

11.9   The Contractor at all times shall keep the premises free from accumulation of waste materials or rubbish caused by his operations. At the completion of the Work he shall remove all his waste materials and rubbish from and about the Project as well as his tools, construction equipment, machinery and surplus materials, and shall clean all glass surfaces and shall leave the Work "broom clean" or its equivalent, except as otherwise specified.

11.10   The Contractor shall indemnify and hold harmless the Owner and the Archiect and their agents and employees from and against all claims, damages, losses and expenses including attorney's fees arising out of or resulting from the performance of the Work, provided that any such claim, damage, loss or expense (a) is attributable to bodily injury, sickness, disease or death, or to injury to or destruction of tangible property (other than the Work itself) including the loss of use resulting therefrom, and (b) is caused in whole or in part by any negligent act or omission of the Contractor, any Subcontractor, anyone directly or indirectly employed by any of them or anyone for whose acts any of them may be liable, regardless of whether or not it is caused in part by a party indemnified hereunder. In any and all claims against the Owner or the Architect or any of their agents or employees by any employee of the Contractor, any Subcontractor, anyone directly or indirectly employed by any of them or anyone for whose acts any of them may be liable, the indemnification obligation under this Paragraph 11.10 shall not be limited in any way by any limitation on the amount or type of damages, compensation or benefits payable by or for the Contractor or any Subcontractor under workmen's compensation acts, disability benefit acts or other employee benefit acts. The obligations of the Contractor under this Paragraph 11.10 shall not extend to the liability of the Architect, his agents or employees arising out of (1) the preparation or approval of maps, drawings, opinions, reports, surveys, Change Orders, designs or specifications, or (2) the giving of or the failure to give directions or instructions by the Architect, his agents or employees provided such giving or failure to give is the primary cause of the injury or damage.

ARTICLE 12

SUBCONTRACTS

12.1 A Subcontractor is a person who has a direct contract with the Contractor to perform any of the Work at the site.

12.2 Prior to the award of the Contract the Contractor shall furnish to the Architect in writing a list of the names of Subcontractors proposed for the principal portions of the Work. The Contractor shall not employ any Subcontractor to whom the Architect or the Owner may have a reasonable objection. The Contractor shall not be required to employ any Subcontractor to whom he has a reasonable objection. Contracts between the Contractor and the Subcontractor shall be in accordance with the terms of this Agreement and shall include the General Conditions of this Agreement insofar as applicable.

ARTICLE 13

SEPARATE CONTRACTS

The Owner has the right to let other contracts in connection with the Work and the Contractor shall properly cooperate with any such other contractors.

ARTICLE 14

ROYALTIES AND PATENTS

The Contractor shall pay all royalties and license fees. The Contractor shall defend all suits or claims for infringement or any patent rights and shall save the Owner harmless from loss on account thereof.

ARTICLE 15

ARBITRATION

All claims or disputes arising out of this Contract or the breach thereof shall be decided by arbitration in accordance with the Construction Industry Arbitration Rules of the American Arbitration Association then obtaining unless the parties mutually agree otherwise. Notice of the demand for arbitration shall be filed in writing with the other party to the Contract and with the American Arbitration Association and shall be made within a reasonable time after the dispute has arisen.

ARTICLE 16

TIME

16.1   All time limits stated in the Contract Documents are of the essence of the Contract.

16.2   If the Contractor is delayed at any time in the progress of the Work by changes ordered in the Work, by labor disputes, fire, unusual delay in transportation, unavoidable casualties, causes beyond the Contractor's control, or by any cause which the Architect may determine justifies the delay, then the Contract Time shall be extended by Change Order for such reasonable time as the Architect may determine.

ARTICLE 17

PAYMENTS

17.1   Payments shall be made as provided in Article 5 of this Agreement.

17.2   Payments may be withheld on account of (1) defective Work not remedied, (2) claims filed, (3) failure of the Contractor to make payments properly to Subcontractors or for labor, materials, or equipment, (4) damage to another contractor, or (5) unsatisfactory prosecution of the Work by the Contractor.

17.3   Final payment shall not be due until the Contractor has delivered to the Owner a complete release of all liens arising out of this Contract or receipts in full covering all labor, materials and equipment for which a lien could be filed, or a bond satisfactory to the Owner idemnifying him against any lien.

17.4   The making of final payment shall constitute a waiver of all claims by the Owner except those arising from (1) unsettled liens, (2) faulty or defective Work appearing after Substantial Completion, (3) failure of the Work to comply with the requirements of the Contract Documents, or (4) terms of any special guarantees required by the Contract Documents. The acceptance of final payment shall constitute a waiver of all claims by the Contractor except those previously made in writing and still unsettled.

ARTICLE 18

PROTECTION OF PERSONS AND PROPERTY

The Contractor shall be responsible for initiating, maintaining, and supervising all safety precautions and programs in connection with the Work. He shall take all reasonable precautions for the safety of, and shall provide all reasonable protection to prevent damage, injury or loss to (1) all employees on the Work and other persons who may be affected thereby, (2) all the Work and all materials and equipment to be incorporated therein, and (3) other property at the site or adjacent thereto. He shall comply with all applicable laws, ordinances, rules, regulations and orders of any public authority having jurisdiction for the safety of persons or property or to protect them from damage, injury or loss. All damage or loss to any property caused in whole or in part by the Contractor, any Subcontractor, any Sub-subcontractor or anyone directly or indirectly employed by any of them, or by anyone for whose acts any of them may be liable, shall be remedied by the Contractor, except damage or loss attributable to faulty Drawings or Specifications or to the acts or omissions of the Owner or Architect or anyone employed by either of them or for whose acts either of them may be liable but which are not attributable to the fault or negligence of the Contractor.

ARTICLE 19

CONTRACTOR'S LIABILITY INSURANCE

The Contractor shall purchase and maintain such insurance as will protect him from claims under workmen's compensation acts and other employee benefit acts, from claims for damages because of bodily injury, including death, and from claims for damages to property which may arise out of or result from the Contractor's operations under this Contract, whether such operations be by himself or by any Subcontractor or anyone directly or indirectly employed by any of them. This insurance shall be written for not less than any limits of liability specified as part of this Contract, or required by law, whichever is the greater, and shall include contractual liability insurance as applicable to the Contractor's obligations under Paragraph 11.10. Certificates of such insurance shall be filed with the Owner.

ARTICLE 20

OWNER'S LIABILITY INSURANCE

The Owner shall be responsible for purchasing and maintaining his own liability insurance and, at his option, may maintain such insurance as will protect him against claims which may arise from operations under the Contract.

ARTICLE 21

PROPERTY INSURANCE

21.1   Unless otherwise provided, the Owner shall purchase and maintain property insurance upon the entire Work at the site to the full insurable value thereof. This insurance shall include the interests of the Owner, the Contractor, Subcontractors and Sub-subcontractors in the Work and shall insure against the perils of Fire, Extended Coverage, Vandalism and Malicious Mischief.

21.2   Any insured loss is to be adjusted with the Owner and made payable to the Owner as trustee for the insureds, as their interests may appear, subject to the requirements of any mortgagee clause.

21.3   The Owner shall file a copy of all policies with the Contractor prior to the commencement of the Work.

21.4   The Owner and Contractor waive all rights against each other for damages caused by fire or other perils to the extent covered by insurance provided under this paragraph. The Contractor shall require similar waivers by Subcontractors and Sub-subcontractors.

ARTICLE 22

CHANGES IN THE WORK

22.1   The Owner without invalidating the Contract may order Changes in the Work consisting of additions, deletions, or modifications, the Contract Sum and the Contract Time being adjusted accordingly. All such Changes in the Work shall be authorized by written Change Order signed by the Owner or the Architect as his duly authorized agent.

22.2   The Contract Sum and the Contract Time may be changed only by Change Order.

22.3   The cost or credit to the Owner from a Change in the Work shall be determined by mutual agreement before executing the Work involved.

ARTICLE 23

CORRECTION OF WORK

The Contractor shall correct any Work that fails to conform to the requirements of the Contract Documents where such failure to conform appears during the progress of the Work, and shall remedy any defects due to faulty materials, equipment or workmanship which appear within a period of one year from the Date of Substantial Completion of the Contract or within such longer period of time as may be prescribed by law or by the terms of any applicable special guarantee required by the Contract Documents. The provisions of this Article 23 apply to Work done by Subcontractors as well as to Work done by direct employees of the Contractor.

ARTICLE 24

TERMINATION BY THE CONTRACTOR

If the Architect fails to issue a Certificate of Payment for a period of thirty days through no fault of the Contractor, or if the Owner fails to make payment thereon for a period of thirty days, the Contractor may, upon seven days' written notice to the Owner and the Architect, terminate the Contract and recover from the Owner payment for all Work executed and for any proven loss sustained upon any materials, equipment, tools, and construction equipment and machinery including reasonable profit and damages.

ARTICLE 25

TERMINATION BY THE OWNER

If the Contractor defaults or neglects to carry out the Work in accordance with the Contract Documents or fails to perform any provision of the Contract, the Owner may, after seven days' written notice to the Contractor and without prejudice to any other remedy he may have, make good such deficiencies and may deduct the cost thereof from the payment then or thereafter due the Contractor or, at his option, may terminate the Contract and take possession of the site and of all materials, equipment, tools, and construction equipment and machinery thereon owned by the Contractor and may finish the Work by whatever method he may deem expedient, and if the unpaid balance of the Contract Sum exceeds the expense of finishing the Work, such excess shall be paid to the Contractor, but if such expense exceeds such unpaid balance, the Contractor shall pay the difference to the Owner.

This Agreement executed the day and year first written above.

OWNER _____ CONTRACTOR _____

SUBCONTRACT FORM

(Developed as a guide by The Associated General Contractors of America, The National Electrical Contractors Association, The Mechanical Contractors Association of America, The Sheet Metal and Air Conditioning Contractors National Association and the National Association of Plumbing—Heating—Cooling Contractors © 1966 by the Associated General Contractors of America and the Council of Mechanical Specialty Contracting Industries, Inc.)

THIS AGREEMENT made this   day of   in the year Nineteen
Hundred and   ,   by and between

                                                                     hereinafter
called the Subcontractor and
hereinafter called the Contractor.
WITNESSETH, That the Subcontractor and Contractor for the consideration hereinafter named agree as follows:

ARTICLE I. The Subcontractor agrees to furnish all material and perform all work as described in Article II hereof for
                                               (Here name the project.)
for
                      (Here name the Contractor.)

at
     (Here insert the location of the work and name of Owner.)

in accordance with this Agreement, the Agreement between the Owner and Contractor, and in accordance with the General Conditions of the Contract, Supplementary General Conditions, the Drawings and Specifications and addenda prepared by

hereinafter called the Architect or Owner's authorized agent, all of which documents, signed by the parties thereto or identified by the Architect or Owner's authorized agent, form a part of a Contract between the Contractor and the Owner dated   , 19   , and hereby become a part of this contract, and herein referred to as the Contract Documents, and shall be made available to the Subcontractor upon his request prior to and at anytime subsequent to signing this Subcontract.

ARTICLE II. The Subcontractor and the Contractor agree that the materials and equipment to be furnished and work to be done by the Subcontractor are: (Here insert a precise description of the work, preferably by reference to the numbers of the drawings and the pages of the specifications including addenda and accepted alternates.)

ARTICLE III. Time is of the essence and the Subcontractor agrees to commence and to complete the work as described in Article II as follows: (Here insert any information pertaining to the method of notification for commencement of work, starting and completion dates, or duration, and any liquidated damage requirements.)

(a) No extension of time of this contract will be recognized without the written consent of the Contractor which consent shall not be withheld unreasonably consistent with Article X-4 of this Contract, subject to the arbitration provisions herein provided.

ARTICLE IV. The Contractor agrees to pay the Subcontractor for the performance of this work the sum of   ($)
in current funds, subject to additions and deductions for changes as may be agreed upon in writing, and to make monthly payments on account thereof in accordance with Article X,

Sections 20-23 inclusive. (Here insert additional details—unit prices, etc., payment procedure including date of monthly applications for payment, payment procedure if other than on a monthly basis, consideration of materials safely and suitably stored at the site or at some other location agreed upon in writing by the parties—and any provisions made for limiting or reducing the amount retained after the work reaches a certain stage of completion which should be consistent with the Contract Documents.)

ARTICLE V. Final payment shall be due when the work described in this contract is fully completed and performed in accordance with the Contract Documents, and payment to be consistent with Article IV and Article X, Sections 18, 20-23 inclusive of this contract.

Before issuance of the final payment the Subcontractor if required shall submit evidence satisfactory to the Contractor that all payrolls, material bills, and all known indebtedness connected with the Subcontractor's work have been satisfied.

ARTICLE VI. Performance and Payment Bonds.
(Here insert any requirement for the furnishing of performance and payment bonds.)

ARTICLE VII. Temporary Site Facilities.
(Here insert any requirements and terms concerning temporary site facilities, i.e., storage, sheds, water, heat, light, power, toilets, hoists, elevators, scaffolding, cold weather protection, ventilating, pumps, watchman service, etc.)

ARTICLE VIII. INSURANCE'

Unless otherwise provided herein, the Subcontractor shall have a direct liability for the acts of his employees and agents for which he is legally responsible, and the Subcontractor shall not be required to assume the liability for the acts of any others.

Prior to starting work the insurance required to be furnished shall be obtained from a responsible company or companies to provide proper and adequate coverage and satisfactory evidence will be furnished to the Contractor that the Subcontractor has complied with the requirements as stated in this Section.

(Here insert any insurance requirements and Subcontractor's responsibility for obtaining, maintaining and paying for necessary insurance, not less than limits as may be specified in the Contract Documents or required by laws. This to include fire insurance and extended coverage, consideration of public liability, property damage, employer's liability, and workmen's compensation insurance for the Subcontractor and his employees. The insertion should provide the agreement of the Contractor and the Subcontractor on subrogation waivers, provision for notice of cancellation, allocation of insurance proceeds, and other aspects of insurance.)

(It is recommended that the AGC Insurance and Bonds Checklist (AGC Form No. 29) be referred to as a guide for other insurance coverages.)

ARTICLE IX. Job Conditions.

(Here insert any applicable arrangements and necessary cooperation concerning labor matters for the project.)

ARTICLE X. In addition to the foregoing provisions the parties also agree:

That the Subcontractor shall:

(1)   Be bound to the Contractor by the terms of the Contractor Documents and this Agreement, and assume toward the Contractor all the obligations and responsibilities that the Contractor, by those documents, assumes toward the Owner, as applicable to this

Subcontract. (a) Not disciminate against any employee or applicant for employment because of race, creed, color, or national origin.

(2)　Submit to the Contractor applications for payment at such times as stipulated in Article IV so as to enable the Contractor to apply for payment.

If payments are made on valuations of work done, the Subcontractor shall, before the first application, submit to the Contractor a schedule of values of the various parts of the work, aggregating the total sum of the Contract made out in such detail as the Subcontractor and Contractor may agree upon, or as required by the Owner, and, if required, supported by such evidence as to its correctness as the Contractor may direct. This schedule, when approved by the Contractor, shall be used as a basis for Certificates for Payment, unless it be found to be in error. In applying for payment the Subcontractor shall submit a statement based upon this schedule.

If payments are made on account of materials not incorporated in the work but delivered and suitably stored at the site, or at some other location agreed upon in writing, such payments shall be in accordance with the terms and conditions of the Contract Documents.

(3)　Pay for all materials and labor used in, or in connection with, the performance of this contract, through the period covered by previous payments received from the Contractor, and furnish satisfactory evidence when requested by the Contractor, to verify compliance with the above requirements.

(4)　Make all claims for extras, for extensions of time and for damage for delays or otherwise, promptly to the Contractor consistent with the Contract Documents.

(5)　Take necessary precaution to properly protect the finished work of other trades.

(6)　Keep the building and premises clean at all times of debris arising out of the operation of this subcontract. The Subcontractor shall not be held responsible for unclean conditions caused by other contractors or subcontractors, unless otherwise provided for.

(7)　Comply with all statutory and/or contractual safety requirements applying to his work and/or initiated by the Contractor, and shall report within 3 days to the Contractor any injury to the Subcontractor's employees at the site of the project.

(8)　(a) Not assign this subcontract or any amounts due or to become due thereunder without the written consent of the contractor. (b) Nor subcontract the whole of this subcontract without the written consent of the contractor. (C) Nor further subcontract portions of this subcontract without written notification to the contractor when such notification is requested by the contractor.

(9)　Guarantee his work against all defects of materials and/or workmanship as called for in the plans, specifications and addenda, or if no guarantee is called for, then for a period of one year from the dates of partial or total acceptance of the Subcontractor's work by the Owner.

(10)　And does hereby agree that if the Subcontractor should neglect to prosecute the work diligently and properly or fail to perform any provision of this contract, the Contractor, after three days written notice to the Subcontractor, may, without prejudice to any other remedy he may have, make good such deficiencies and may deduct the cost thereof from the payment then or thereafter due the Subcontractor, provided, however, that if such action is based upon faulty workmanship the Architect or Owner's authorized agent, shall first have determined that the workmanship and/or materials is defective.

(11)　And does hereby agree that the Contractor's equipment will be available to the Subcontractor only at the Contractor's discretion and on mutually satisfactory terms.

(12) Furnish periodic progress reports of the work as mutually agreed including the progress of materials or equipment under this Agreement that may be in the course of preparation or manufacture.

(13) Make any and all changes or deviations from the original plans and specifications without nullifying the original contract when specifically ordered to do so in writing by the Contractor. The Subcontractor prior to the commencement of this revised work, shall submit promptly to the Contractor written copies of the cost or credit proposal for such revised work in a manner consistent with the Contract Documents.

(14) Cooperate with the Contractor and other Subcontractors whose work might interfere with the Subcontractor's work and to participate in the preparation of coordinated drawings in areas of congestion as required by the Contract Documents, specifically noting and advising the Contractor of any such interference.

(15) Cooperate with the Contractor in scheduling his work so as not conflict or interfere with the work of others. To promptly submit shop drawings, drawings, and samples, as required in order to carry on said work efficiently and at speed that will not cause delay in the progress of the Contractor's work or other branches of the work carried on by other Subcontractors.

(16) Comply with all Federal, State and local laws and ordinances applying to the building or structure and to comply and give adequate notices relating to the work to proper authorities and to secure and pay for all necessary licenses or permits to carry on the work as described in the Contract Documents as applicable to this Subcontract.

(17) Comply with Federal, State and local tax laws, Social Security laws and Unemployment Compensation laws and Workmen's Compensation Laws insofar as applicable to the performance of his subcontract.

(18) And does hereby agree that all work shall be done subject to the final approval of the Architect or Owner's authorized agent, and his decision in matters relating to artistic effect shall be final, if within the terms of the Contract Documents.

That the Contractor shall—

(19) Be bound to the Subcontractor by all the obligations that the Owner assumes to the Contractor under the Contract Documents and by all the provisions thereof affording remedies and redress to the Contractor from the Owner insofar as applicable to this Subcontract.

(20) Pay the Subcontractor within seven days, unless otherwise provided in the Contract Documents, upon the payment of certificates issued under the Contractor's schedule of values, or as described in Article IV herein. The amount of the payment shall be equal to the percentage of completion certified by the Owner or his authorized agent for the work of this Subcontractor applied to the amount set forth under Article IV and allowed to the Contractor on account of the Subcontractor's work to the extent of the Subcontractor's interest therein.

(21) Permit the Subcontractor ot obtain direct from the Architect or Owner's authorized agent, evidence of percentages of completion certified on his account.

(22) Pay the Subcontractor on demand for his work and/or materials as far as executed and fixed in place, less the retained percentage, at the time the payment should be made to the Subcontractor if the Architect or Owner's authorized agent fails to issue the certificate for any fault of the Contractor and not the fault of the Subcontractor or as otherwise provided herein.

(23)　And does hereby agree that the failure to make payments to the Subcontractor as herein provided for any cause not the fault of the Subcontractor, within 7 days from the Contractor's receipt of payment or from time payment should be made as provided in Article X, Section 22, or maturity, then the Subcontractor may upon 7 days written notice to the Contractor stop work without prejudice to any other remedy he may have.

(24)　Not issue or give any instructions, order or directions directly to employees or workmen of the Subcontractor other than to the persons designated as the authorized representative(s) of the Subcontractor.

(25)　Make no demand for liquidated damages in any sum in excess of such amount as may be specifically named in the subcontract, provided, however, no liquidated damages shall be assessed for delays or causes attributable to other Subcontractors or arising outside the scope of this Subcontract.

(26)　And does hereby agree that no claim for services rendered or materials furnished by the Contractor to the Subcontractor shall be valid unless written notice thereof is given by the Contractor to the Subcontractor during the first ten days of the calendar month following that in which the claim originated.

(27)　Give the Subcontractor an opportunity to be present and to submit evidence in any arbitration involving his rights.

(28)　Name as arbitor under arbitration proceedings as provided in the General Conditions the person nominated by the Subcontractor, if the sole cause of dispute is the work, materials, rights or responsibilities of the Subcontractor; or if, of the Subcontractor and any other Subcontractor jointly, to name as such arbitrator the person upon whom they agree.

That the Contractor and the Subcontractor agree—

(29)　That in the matter of arbitration, their rights and obligations and all procedure shall be analogous to those set forth in the Contract Documents provided, however, that a decision by the Architect or owner's authorized agent, shall not be a condition precedent to arbitration.

(30)　This subcontract is solely for the benefit of the signatories hereto.

ARTICLE XI.

IN WITNESS WHEREOF the parties hereto have executed this Agreement under seal, the day and year first above written.

Attest:	Subcontractor	
(Seal)	By	(Title)
Attest:	Contractor	
(Seal)	By	(Title)

Chapter 9

Cost-Plus-Fee Construction Contracts

Underlying Principles

The basic principle underlying cost-plus-fee contracts is that the contractor will be reimbursed by the owner for all costs attributable to the project and, in addition, he will be paid a fee for his services. Thus it is intended that the contractor be assured of a profit on his operations. This assurance eliminates divergence of financial interests, and, when the fee is a fixed sum, the relationship of contractor to owner tends to become a professional one. The contractor exercises his skill in the techniques of construction, procurement of labor and materials, and management for the owner's benefit. His selection for the work is based on his past record of experience, skill, and dependability. The owner, on the other hand, by electing to build the project under a cost-plus-fee arrangement tacitly indicates that factors other than lowest possible cost are paramount. Among these may be the desire for high quality construction or an urgency that requires earliest possible start ahead of the time complete drawings and specifications will be available.

Cost-plus-fee contracts are essentially open-ended, since the owner cannot know accurately what the construction cost will be until the project is completed. If, however, the contractor is selected early in the design stage so that he can work closely with the architect and consulting engineers, his skill in estimating and his knowledge of the relative economics of different construction methods should produce estimates that reflect the final cost within reasonably narrow limits.

The amount of the fee may be established as a percentage of the actual cost of construction or as a fixed fee agreed upon in advance. The first arrangement makes the fee open-ended as well as the cost of the work; it is not recommended except where the nature of a project or portion thereof is such that determination of scope prior to construction is impracticable. This is sometimes the case in alteration and repair work, underpinning, and other special operations. Cost-plus-percentage agreements are prohibited by law in most public works construction.

Under cost-plus-fixed-fee (CPFF) contracts there is no incentive for the contractor to inflate construction costs. On the other hand, neither is there an inherent one to encourage keeping them down. Nevertheless, many building contractors have achieved a reputation for integrity, efficiency, and economy under CPFF operations by meeting their target estimates. Cost-plus-fixed-fee contracts should not be entered into lightly by an owner, but they have many desirable features once the possibility of runaway costs has been restricted by selection of a competent and reliable contractor.

The CPFF Agreement

As with lump-sum construction contracts, the CPFF Agreement is the instrument that welds all the Contract Documents into a single contract. Standard contract forms are of less general applicability in relation to cost-plus contracts because of the diversity of possible arrangements. The two-document pattern for Agreement and Conditions of the Contract is fully as advantageous for CPFF as for lump-sum contracts, but standardized General Conditions developed for lump-sum contracts require some modification for use with CPFF Agreements. This is particularly true of articles relating to contract sum, subcontracts, payments to the contractor, and changes in the work. Although detailed provisions are subject to considerable variation, all CPFF Agreements contain articles covering the following topics:

Identification of parties and project
Identification of contract documents
Contract time and liquidated damages
Reimbursable costs
Nonreimbursable costs
Contractor's fee
Changes in the work
Subcontracts
Payments to the contractor
Accounting and cost control
Miscellaneous provisions

A consideration of each one follows. It is recommended that the reader compare the discussions with corresponding articles in the CPFF contract form reproduced at the end of the chapter.

Identification of Parties and Project

The same information is required here as that described under the similar headings for lump-sum Agreements, and again the date appearing in the preamble is the effective date of the contract. The status of both owner and contractor with respect to form of business organization, that is, individual, partnership, joint venture, or corporation, should be stated, together with the legal address of each party. The architect should also be identified by name, although, of course, he is not a party to the contract.

For purposes of the Agreement the project may be identified by the descriptive title used on the Drawings and Specifications, with more detailed information concerning location of the site and location of the project shown by the site plan.

Identification of Contract Documents

Special consideration must be given to this article when complete plans and project specifications are not available at the time the Agreement is executed. This situation occurs frequently since one of the merits of a CPFF contract is the opportunity afforded for early start of construction. The Conditions of the Contract present no special problem in this regard since they may be fully prepared in advance, but some recognition should be made therein and in the Agreement that the Contract Documents enumerated at the time of execution will be amplified by ". . . subsequent drawings and written instructions which shall be within the original intent of the existing Drawings and Specifications." Such a clause affords protection to the contractor against changes in scope without renegotiation of the fixed fee and protection to the owner against claims for change orders that imply change in scope.

The scope of work to be performed can generally be estimated with sufficient accuracy from the definitive design drawings and outline specifications prepared during the design development phase of architectural and engineering services. Consequently, these documents may be cited in the Agreement as defining the intended scope of the project which will be developed more precisely by the "subsequent plans and written instructions." This is not quite the same provision as that noted under lump-sum Agreements where "all Modifications" (change orders, supplemental amendments, etc.) made subsequent to execution become part of the contract. In that case, the Drawings, Specifications, and Addenda issued before execution of the contract presumably defined the intended scope, with subsequent Modifications providing a mechanism for changes including additional compensation to the contractor.

Although change orders involving change in scope may be issued under CPFF contracts, that is not the function of the "subsequent plans and written instructions" discussed; these simply provide additional information for further definition of work contemplated in the original scope. This matter is considered further in the articles on contractor's fee and changes in the work.

Aside from this special consideration for handling nonavailability of complete plans and project specifications at the time the contract is executed, the clause making all of the documents an integrated contract is similar to the one used in lump-sum Agreements. It may then be phrased as follows:

"The Contract Documents consist of this Agreement, Conditions of the Contract, Drawings, Specifications, and any subsequent drawings and written instructions which shall be within the original intent of the existing Drawings, Specifications and other documents enumerated below. These form the contract, and are all as fully a part thereof as if attached to this Agreement or repeated herein."

The detailed enumeration is often placed at the end of the Agreement because of its uncertain length, where "below" is replaced by a suitable article number reference. As with lump-sum Agreements, each document for which a standard form is not employed should be listed to indicate the number of pages involved. This should always be done for each division of the Specifications and, of course, sheet numbers must be given to identify the several sets of drawings, that is, architectural, structural, mechanical, etc.

Contract Time

Since the requirement for an early occupancy date is often the principal reason for negotiating a cost-plus-fixed-fee contract, special consideration should be given to the time of commencement and completion. In addition to the advantages tages of selecting the contractor early in the design stage there is the added one that development of working drawings and related specification sections can be keyed to a projected construction schedule. The time of commencement could then be designated in the Agreement as the date working drawings and relevant specifications for the first phase of the work are issued by the architect and the contract time and date of substantial completion established. If there is insufficient information at that stage to foresee the length of time required for completion, some other provision for establishing the contract time should be devised; otherwise no basis may exist for initiating termination proceedings in event of unreasonable delay by the contractor in carrying on the work.

Liquidated damages provisions may be of little value under CPFF contracts when the contract time cannot be determined with reasonable accuracy before execution of the Agreement. Probably the best safeguard under such conditions is the selection of a contractor who has an established reputation for meeting his deadlines. Even if unforeseen foundation conditions are encountered with consequent delay in getting the project "out of the ground," an early start plus time gained by elimination of a formal bidding period may do more to assure completion on time than a deferred start accompanied by liquidated damages provisions. Where provision for liquidated damages forms part of a CPFF Agreement, it should be stipulated that in event they are assessed they constitute a direct charge to the contractor, presumably from his fixed fee. Aside from this point, the discussion of liquidated damages presented in Chapter 8 substantially covers the situation with respect to CPFF contracts and will not be repeated here.

Reimbursable Costs

Since CPFF contracts are predicated on the contractor being reimbursed for all costs of the work and being paid in addition a fee for his services, it is necessary to differentiate between reimbursable costs and those costs deemed to be covered by the fixed fee. This would not pose much of a problem if the contractor's profit were the only element in his fee, but this is not the case. One way to approach allocation of costs is to define precisely items covered by the fixed fee and then declare that all other expenditures made by the contractor in good faith constitute the "cost of the work." This view is not customary, however, and most CPFF Agreements specify in considerable detail the costs that are reimbursable and also identify a few that definitely are not. The costs that are not reimbursable are considered in the next article after examining here items generally recognized to constitute the "cost" portion of a cost-plus-fee contract.

It should be noted at this point that the total cost of the work involves both direct and indirect costs, that is, those that can be itemized as payment for specific items of material, equipment, labor, services, etc., and those overhead costs of the contractor's central organization that can only be allocated proportionately to an individual project. These indirect costs are just as real as the direct ones, but because of the difficulty of evaluating them, they are frequently considered as expenses to be paid by the contractor from his fee.

Any listing of direct costs which are reimbursable must be treated as a guide only since every project has some particular characteristics. Obviously the provisions relating to reimbursable and nonreimbursable costs lie at the heart of CPFF Agreements and can produce no end of difficulty if not carefully drawn. The following items of cost are among those listed as reimbursable costs in most CPFF Agreements.

WAGES

These cover all labor directly employed by the contractor either at the site or away from the site engaged in prefabrication, transportation, warehousing, and similar functions essential to the project. The cost of this labor includes payroll taxes, unemployment insurance, and other welfare benefits in accordance with prevailing labor agreements.

SALARIES

These are limited to salaries of the contractor's employees assigned to the field office and cover those engaged in inspection and expediting production of materials, components, and equipment away from the site to the extent their activity relates to the project. In computing salary expense the cost of mandatory and customary employee benefits is included.

SUBCONTRACTS

In addition to the subcontract amounts, the total costs here include premiums for performance and payment bonds.

MATERIALS AND EQUIPMENT

The cost of all materials, supplies, and equipment includes associated expenses for transportation, inspection, storage, and any applicable sales taxes.

TRAVEL EXPENSE

This includes the cost of transportation, hotel, and other subsistence expense incurred by members of the contractor's organization when traveling on project business.

FIELD FACILITIES

The costs involved in this connection include construction, furnishing, maintenance, and subsequent removal of the field office, storage and tool sheds, and other temporary structures, with any salvage value on completion of the project reverting to the owner. Also included under this heading are costs of access roads and field utilities required during construction operations such as electricity, water, fuel, and telephone.

PLANT AND EQUIPMENT

Determination of costs associated with construction plant and equipment presents one of the most troublesome aspects of cost-plus-fee contracts. Bulldozers, cranes, scaffolding, and hoists, among other items of equipment, wear out or deteriorate over the course of several jobs and consequently their owners must provide for their upkeep, repair and eventual replacement. The difficulty arises in the effort to translate these factors into equitable costs to be reimbursed under CPFF contracts. Of course, the same basic problem exists under lump-sum contracts, but there it is resolved by the contractor without involving the building owner except as these considerations affect the level of bids received. However, the open-end character of cost-plus contracts requires that procedures for handling plant and equipment costs be specified clearly if misunderstandings and disputes are to be minimized. The customary approach is to procure these items through rental agreements whether rented from the contractor or others.

Rental rates are frequently determined in accordance with the procedure recommended by the Associated General Contractors of America which is based on a flat rate for each type of equipment. Such rates are customarily predicated on ownership expense only (depreciation, major repairs, taxes, insurance, etc.), since operating expenses are charged as direct costs against the project. No allowance for profit to the contractor should be included in rental rates whether the equipment is rented from him or from third parties, all of the contractor's profit being covered by the fixed fee. It should be noted, however, that in rentals from third parties there will be a profit factor for the owner of the equipment. Since this is passed along to the project owner as a reimbursable cost, it is beneficial to have an understanding beforehand concerning the items of equipment to be rented from third parties and the rates to be paid.

In addition to rental charges, the contractor is also reimbursed for costs associated with delivery, transportation, installation, and removal of plant and equipment items. More troublesome is the matter of repairs where a distinction has to be made between normal operating maintenance and repairs and replacements attributable to accidents of use at the site. Rental agreements vary in this respect, but their provisions should be scrutinized so there is no misunderstanding concerning which repair and maintenance costs are to be reimbursed by the owner.[1]

On large projects it may be economical to have the contractor purchase certain items of plant and equipment for the owner as a reimbursable cost. This procedure has the advantage of providing new equipment which should promote efficiency of job operations and also removes any ambiguity with respect to responsibility for maintenance, repair, and overhaul. When the project is completed, such equipment can be disposed of, with any resale or salvage value reverting to the owner.

SMALL TOOLS AND EQUIPMENT

The costs involved here cover small tools not customarily furnished by workmen, canvas tarpaulins, or other sheeting-type covers and enclosing materials, and similar items for which rental procedures are impracticable. Many items supplied under this heading are used up during the course of the job and should be considered expendable. The simplest procedure is to declare their costs reimbursable and then provide for the contractor to take over any that remain at the end of the work at an agreed valuation. It is sometimes proposed to cover items of this nature by a lump sum, but this is not really a simplification, since

[1]For a more detailed discussion of equipment expense from the contractor's point of view see R. H. Clough, *Construction Contracting* (New York: Wiley, 1960), pp. 222–226.

the question of whether any particular item falls under the lump sum or constitutes a reimbursable cost has still to be determined.

BONDS AND INSURANCE

This includes premiums for performance and payment bonds if required, and those for all forms of insurance required by law or called for in the Conditions of the Contract, including public liability and property damage, workmen's compensation, project loss or damage, and any other type of insurance required by the owner or customarily carried by the contractor when prosecuting the work of a specific project. Of course, the owner may elect to take out certain insurance coverage himself as discussed in Chapter 6 and this should be made clear in the Conditions of the Contract. The purpose of this clause in the Agreement is simply to allow as reimbursable costs the premiums for insurance which the contractor is required to carry.

LOSSES

Reimbursement of the contractor under this provision is to cover losses and expenses due to flood, earthquake, riot, and similar causes not compensated by insurance or otherwise, provided that they were not occasioned through fault or neglect of the contractor. Where reconstruction is carried out by the contractor after such loss, a fee proportionate to the original fixed fee should be allowed as a reimbursable cost.

PERMITS, ROYALTIES, AND PATENTS

Fees for construction permits required of the contractor constitute reimbursable items of cost as do license fees and royalties covering use of patented systems and methods of construction, and any damages that may be awarded for patent infringement as well as the costs of defending law suits in connection therewith.

COMMUNICATION SERVICES

This includes the net cost of telephone calls, telegrams, cablegrams, and postage.

OTHER EXPENSES

In order to provide flexibility for unforeseen expenses, it is customary to include a clause providing that other expenses and charges pertaining to the project will be allowed as reimbursable costs if the owner has given written approval before such costs are incurred by the contractor.

Nonreimbursable Costs

Since it is intended under a CPFF contract that the contractor be reimbursed for *all* costs incurred in construction of the project, this category of nonreimbursable costs relates to those that are nonreimbursable under the cost provisions of the contract and hence must be accounted for in the fixed fee. In general, these are indirect costs and, as noted earlier, they represent the overhead expenses of the contractor's central organization that can be allocated to an individual project on a proportionate basis only. For example, the cost of operating the contractor's payroll department could be apportioned among the several jobs in progress in proportion to the dollar amounts of the payrolls of the individual projects. This illustration is not an ironclad guide, since on large projects where the scope of operations warrants, a payroll staff can be established in the field office; all of its expenses including salaries of the personnel stationed there then become items of reimbursable cost. However, any supervision that might be exercised over the field payroll staff by the treasurer of the company or the head bookkeeper would remain part of the central office overhead.

It will be apparent that making a rational allocation of overhead costs when several projects are under way concurrently is a difficult task for the contractor. This is especially true with such general expenses as office rent, office insurance, furniture, light, legal services, and advertising, and the salaries of executives and office employees. All of these represent valid costs of doing business and have to be covered in some manner. In general building construction it is often possible for a contractor to determine over a period of time a proportional relationship of overhead expense to direct labor costs or to direct labor plus direct material costs. Where this determination can be validated by auditing procedures acceptable to the owner, a fixed percentage of such direct costs may be allowed as a reimbursable overhead cost. Where this method of handling overhead expense can be applied, accounting procedures are greatly simplified, thereby saving time and expense.

By and large the items identified under nonreimbursable costs constitute a table of contents for the overhead category. Here again any list of such items must be treated as only a guide since the characteristics of individual projects will require some modification. However, the following items are generally among those enumerated as nonreimbursable costs in CPFF Agreements.

CENTRAL ORGANIZATION SALARIES

Among these are the salary of the contractor, if an individual, and those of the company's officers, executives, and managers if organized as a corporation or other form of business. Also included in this category are salaries of purchasing agents, estimators, accountants, and other department heads and their supporting staffs. In fact, the only occasion when staff members (and then not officers of the company) can have their time charged directly to a project occurs when they are away from the office engaged in expediting or inspecting materials and equipment under procurement for the project as noted in the reimbursable costs article.

HEADQUARTERS AND BRANCH OFFICE EXPENSE

This consists of general office overhead including rent, insurance, supporting secretarial personnel, etc., whether incurred by the contractor's central organization at its principal office or at a branch office.[2]

OPERATIONS EXPENSE

It will be apparent from the discussion thus far that certain items of the contractor's overhead can, with more or less difficulty, be segregated and allocated to individual projects. However, there are others that defy any rational breakdown; among these are costs associated with labor relations, general legal counsel, arbitration, advertising and promotional activity, public relations, preparation of subcontracts, budgets, and construction schedules. Almost invariably these are classified as nonreimbursable costs and must be covered by the fixed fee.

OVERTIME PAY

This as well as multiple-shift operations may be allowed, but only when specifically authorized by the owner, unless provided for by an agreed allowance under a CPFF contract with a guaranteed maximum cost.

GENERAL EXCLUSION

Some CPFF Agreements contain a clause that classifies any item not specifically included in the list of reimbursable costs as nonreimbursable. This somewhat one-sided provision no doubt serves as another safeguard offsetting the open-end character of any cost-plus contract; when it is used, however, care should be taken to see that an item covering additional expenses "approved by the owner before being incurred" appears in the reimbursable cost article. As noted earlier this provides necessary flexibility for handling unforeseen developments and still retains a check on runaway costs.

Contractor's Fee

There are two principal components of the fixed fee: the contractor's profit and that portion of his expense of doing business that cannot be allocated to the project as a reimbursable cost. Even on contracts where reimbursement for certain indirect costs is allowed as reimbursable overhead, there will always be some indirect costs not so covered. This is particularly true on federal

---

[2]Not to be confused with a project field office. Many large contracting firms have a principal or headquarters office with branch offices located in different parts of the country. Both principal and branch offices are considered part of the contractor's central organization in this discussion.

government contracts where certain elements of overhead cost are always considered as nonreimbursable and consequently must be paid out of the fixed fee. In effect, the contractor estimates a lump sum to cover the cost of all nonreimbursable overhead items, and this plus the amount allowed for profit constitutes the fixed fee.

The amount of the fee should be based on considerations such as character and scope of the project, time available for construction, estimated construction cost, and the proportion of work to be subcontracted. A close appraisal of the scope is of critical significance, since change in scope constitutes the only basis for a change in the fixed fee. This is true regardless of the accuracy of the cost estimate. For example, a rise in the price of structural steel would cause an increase in the cost of the building frame but would not involve a change in project scope. However, the addition of more floors than originally contemplated would constitute a change in scope as well as a change in cost regardless of any fluctuation in the price of structural steel. It is because the scope of a project and a reasonably accurate approximation of the cost can be determined from preliminary plans and outline specifications that CPFF construction contracts can be negotiated while the project is still in the design stage. At this point the contractor commits himself only to a fixed amount of fee, not to a fixed construction cost.

CPFF WITH GUARANTEED CEILING

Cost-plus-fixed-fee contracts can be (and often are) written with a guaranteed maximum cost. This modification, however, tends to undermine the professional relationship between contractor and owner that is one of the virtues of negotiated CPFF contracts. In addition, nearly complete plans and specifications must be available as a basis for determining the guaranteed maximum cost and this results in loss of much of the potential for an early start.

Unless the guaranteed limit is established from essentially complete construction documents, there is a built-in source of controversy concerning whether work shown on the drawings, as they are developed and issued by the architect-engineer, falls within the scope of the project as understood by the contractor when he agreed to the guaranteed limit. Under such conditions the interests of contractor and owner no longer run in the same direction; it is to the contractor's advantage to press for an increase in the guaranteed limit of cost, which, of course, works to the disadvantage of the owner.

GUARANTEED CEILING WITH PROFIT SHARING

CPFF contracts with guaranteed maximum cost can also be written with a profit-sharing provision to be implemented if the final cost is under the guaranteed maximum. Here the difference between the two figures is shared between owner and contractor on a fifty-fifty basis or other ratio stipulated in the contract. Under such an arrangement the guaranteed maximum cost is subject to adjustment for extra work and change orders in the same manner that adjustments are made in the contract sum on lump-sum contracts.

Many other variations of the cost-plus-fixed-fee contract have been devised with a view to providing additional incentive for the contractor to hold down the cost of the work. Those discussed, however, represent the types most commonly employed in building construction.

Changes in the Work

Under cost-plus-fixed-fee contracts the costs of changes are merged in the final cost of the work. Although the contractor may furnish an estimate of the cost of a proposed change, he is not bound by the estimate as he would be under a lump-sum agreement. Whether an adjustment in the fixed fee is called for depends on the nature of the change, since a change in fee is in order only where the scope of the work has been changed; this is generally understood to mean a material increase or decrease in the amount or character of the work.

When a minor change in scope is involved, the change order should provide for a proportionate adjustment in the fixed fee. If a change in scope is of considerable magnitude, such as when additional buildings or facilities are added to the contract, it is recommended that a supplemental agreement be executed by the owner and contractor to cover the new work. This procedure permits negotiation of a fee based on the character of the change rather than making a strictly proportionate adjustment of the original fee.

When a CPFF contract is written with a guaranteed maximum cost, this sum must, of course, be adjusted to reflect the amounts of all change orders. It is customary to base such adjustments on the contractor's estimate of the cost of proposed changes at the time they are requested by the owner and to make adjustment in the guaranteed maximum when the contractor is authorized to proceed with the change. The change order should also indicate whether a change in contract time is involved.

Subcontracts

One of the merits of the cost-plus-fee type contract is the measure of control it affords in the selection of competent subcontractors. Ordinarily, subcontracts under a CPFF prime contract are drawn on a lump-sum basis. These are usually awarded after competitive bidding by three or more selected subcontractors mutually satisfactory to owner, prime contractor, and architect-engineer. The circumstances surrounding a particular project may make it desirable to negotiate some subcontracts. This situation can arise when it is necessary to select a subcontractor before complete working drawings and specifications for his portion of the work are available. If negotiations with a subcontractor leading to a cost-plus-fee subcontract are to be carried out by the prime contractor, the procedures to be followed should be specified in the prime contract. Since the amounts of subcontracts constitute reimbursable costs, they should be approved by the owner prior to execution of subcontract agreements by the prime contractor. The prime contractor must, of course, have full directing authority and management control over the operations of subcontractors.

Wherever possible lump-sum subcontracts should be employed under CPFF prime contracts. The use of cost-plus-fee subcontract agreements aggravates the already open-end characteristic of the CPFF prime contract and they seldom operate to the owner's economic advantage.

Payments to the Contractor

Unless otherwise provided, payments for materials or services procured under a CPFF contract are advanced by the contractor, who then requests reimbursement from the owner. However, this procedure is often modified in practice, sometimes by the owner advancing operating funds to the contractor or by the owner making payment to the contractor for materials and equipment received on the basis of invoices to be paid upon receipt of funds from the owner. As with lump-sum contracts, the payment process involves the filing of an application for payment by the contractor and the issuing of a certificate for payment by the architect.

APPLICATIONS FOR PAYMENT

The contract should state that on or before a specified day of each month, the contractor shall present to the architect an application for payment covering work completed during the preceding month, together with supporting evidence that the sum requested is due. The application should be based upon a schedule of values as described in the General Conditions (see Payments to the Contractor, Chapter 4) with items adjusted to conform to the current estimated cost of the project.[3] The application should carry a certification that the work indicated has been performed in accordance with the Contract Documents and that all amounts have been paid for items covered by previously issued certificates for payment.

The application should be accompanied by certification of payrolls met during the period, invoices for materials and equipment received, requisitions received from subcontractors for work performed, a statement of reimbursable

---

[3]In CPFF contracts with a guaranteed ceiling, the current value of the guaranteed maximum cost would be used.

expenses paid during the period, and a statement of the proportionate amount of the fixed fee due. Beginning with the second application for payment, the contractor should be required to furnish evidence that he has, in fact, made all payments on account of items covered by the certificate for payment issued for the preceding month.

CERTIFICATES FOR PAYMENT

After review of the contractor's application, a certificate for payment is issued in the full amount approved by the architect, except that an agreed percentage is retained on all lump-sum subcontracts. These retainages are paid on completion of the particular portions of the work involved. Certificates for payment are payable by the owner on issuance by the architect.

The process culminating in the final certificate for payment is much the same as that under lump-sum contracts discussed in Chapter 8. It is customary to provide that the owner shall make final payment within ten days after receipt of the final certificate and that the remaining balance of the fixed fee shall also be paid to the contractor at that time.

Accounting and Cost Control

On projects of limited scope the contractor's vouchers may be accepted for reimbursement by the owner based on the contractor's own records without attempting to verify expenditures for labor and materials by checking at the site. If the work, however, is of considerable magnitude, the owner should employ an auditor versed in construction accounting procedures to maintain a continuous check on all labor and material to be billed as a job cost.

The contractor should be required to keep detailed records and accounts as necessary for proper financial management of the contract according to accounting procedures satisfactory to the owner. At the same time a running record of estimates of cost to complete the various classifications of work shown on the schedule of values should be maintained. This information is of great value to both contractor and owner in appraising the efficiency of the construction operation and the financial status of the project with respect to the estimated cost. It should be emphasized here that auditing services are the responsibility of the owner and should not be undertaken by the architect-engineer unless his firm incorporates such capability under the comprehensive services pattern of practice.

Miscellaneous Provisions

CONTRACTOR'S FINANCIAL RESPONSIBILITY

Although all costs attributable to the project are intended to be paid for by the owner under a CPFF contract, those incurred by the contractor as a result of his own negligence or the negligence of his employees obviously do not fall in this category. Consequently, a clause covering this situation should be incorporated in the contract and be so worded that the costs of making good defective work are to be borne by the contractor. Since work in place and subsequently judged defective by the architect may already have been paid for, provision should be made for the owner to withhold money due the contractor in an amount sufficient to assure that double reimbursement is not made.

DISCOUNTS, REBATES, AND REFUNDS

When the owner advances funds to the contractor to make payment for materials and equipment, the contractor should be required to make these payments so that trade discounts, rebates, and refunds will accrue to the owner. It should also be stipulated that all proceeds from the sale of surplus materials and equipment will accrue to the owner.

TERMINATION OF THE CONTRACT

The discussion of this subject in Chapter 4, as an article of the General Conditions, related principally to lump-sum contracts, but it is also valid for the most part with respect to CPFF contracts. Provision should be made for termination for default or breach on the part of either contractor or owner and for termination at will by the owner without default of the contractor. Termination is a somewhat less complicated matter under cost-plus-fee contracts than under lump-sum agreements. However, when termination is made at the convenience of the owner, the contractor must be compensated for all of his costs arising from the work, including the expense of demobilizing his operations, canceling committed orders for materials and equipment, and similar terminal costs. He will, of course, be entitled to the proportionate amount of the fixed fee earned to date, but the termination provisions of the contract should be specific concerning conditions (if any) under which the contractor is to be paid any part of his anticipated profit on the canceled portion of the work.

Management Contracts

Under this type of cost-plus-fee contract the contractor is in effect the owner's construction manager. Unlike the conventional CPFF contract, where it is customary for the contractor to accomplish a portion of the work with his own forces, a management-type contract contemplates retention of the contractor primarily in a managerial capacity. Under these circumstances it is usual to subcontract the entire work, and the contractor handles all matters pertaining to scheduling, supervision of construction, purchasing of materials and equipment, budget control, and letting of subcontracts. The contract should be specific concerning whether management services only are intended and, obviously, this type of contract should be negotiated with a building construction contractor of acknowledged integrity, administrative ability, and technical skill. When these conditions are fulfilled, management contracts offer many advantages, particularly where the contractor is selected early in the project so that his knowledge of the relative economics of construction methods and costs of materials may be utilized in the design stage.

The management contractor may operate as an independent contractor or as agent of the owner, and the contract should be explicit as to which is intended. Under the independent relationship the contractor remains responsible for the work until completion and acceptance by the owner. All expenditures are made in the contractor's name, although they are generally paid for from funds advanced by the owner. Under the agency relationship all the contractor's actions are taken in the name of the owner, and the owner assumes responsibility for all actions of the contractor. Obviously, the decision regarding which type of relationship shall be implemented is a matter for the owner to consider with his legal counsel.

EXAMPLES OF COST-PLUS-FEE CONTRACTS

As noted earlier, standard forms are of less general applicability with respect to cost-plus-fee contracts because of the diversity of possible arrangements. One form, however, that may serve as a prototype is presented here for illustration. Reproduction is by permission of the American Institute of Architects.

THE AMERICAN INSTITUTE OF ARCHITECTS

AIA Document A111

Standard Form of Agreement Between Owner and Contractor

where the basis of payment is the
COST OF THE WORK PLUS A FEE

Use only with the latest edition of AIA Document A201, General Conditions of the Contract for Construction.

AGREEMENT

made this   day of   in the year of Nineteen
Hundred and

BETWEEN

the Owner, and

the Contractor.

The Owner and the Contractor agree as set forth below.

AIA DOCUMENT A111 • OWNER-CONTRACTOR AGREEMENT • SEPTEMBER 1967 EDITION • AIA® ©1967
THE AMERICAN INSTITUTE OF ARCHITECTS, 1735 NEW YORK AVENUE, N.W., WASHINGTON, D. C. 20006

ARTICLE 1

THE CONTRACT DOCUMENTS

The Contract Documents consist of this Agreement, Conditions of the Contract (General, Supplementary and other Conditions), Drawings, Specifications, all Addenda issued prior to execution of this Agreement and all Modifications issued subsequent thereto. These form the Contract, and all are as fully a part of the Contract as if attached to this Agreement or repeated herein. An enumeration of the Contract Documents appears in Article 17. If anything in the General Conditions is inconsistent with this Agreement, the Agreement shall govern.

ARTICLE 2

THE WORK

The Contractor shall perform all the Work required by the Contract Documents for
(Here insert the caption descriptive of the Work as used on other Contract Documents.)

ARTICLE 3

ARCHITECT

The Architect for this Project is

ARTICLE 4

THE CONTRACTOR'S DUTIES AND STATUS

The Contractor accepts the relationship of trust and confidence established between him and the Owner by this Agreement. He covenants with the Owner to furnish his best skill and judgment and to cooperate with the Architect in furthering the interest of the Owner. He agrees to furnish business administration and superintendence and to use his best efforts to furnish at all times an adequate supply of workmen and materials, and to perform the Work in the best and soundest way and in the most expeditious and economical manner consistent with the interests of the Owner.

ARTICLE 5

TIME OF COMMENCEMENT AND COMPLETION

The Work to be performed under this Contract shall be commenced

and completed
(Here insert any special provisions for liquidated damages relating to failure to complete on time.)

ARTICLE 6

COST OF THE WORK AND GUARANTEED MAXIMUM COST

6.1   The Owner agrees to reimburse the Contractor for the Cost of the Work as defined in Article 9. Such reimbursement shall be in addition to the Contractor's Fee stipulated in Article 7.

6.2   The maximum cost to the Owner, including the Cost of the Work and the Contractor's Fee, is guaranteed not to exceed the sum of

dollars ($); such Guaranteed Maximum Cost shall be increased or decreased for Changes in the Work as provided in Article 8.
(Here insert any provision for distribution of any savings. Delete Paragraph 6.2 if there is no Guaranteed Maximum Cost.)

ARTICLE 7

CONTRACTOR'S FEE

7.1   In consideration of the performance of the Contract, the Owner agrees to pay the Contractor in current funds as compensation for his services a Contractor's Fee as follows:

7.2   For Changes in the Work, the Contractor's Fee shall be adjusted as follows:

7.3   The Contractor shall be paid   per cent (%) of the proportionate amount of his Fee with each progress payment, and the balance of his Fee shall be paid at the time of final payment.

ARTICLE 8

CHANGES IN THE WORK

8.1   The Owner may make Changes in the Work in accordance with Article 12 of the General Conditions insofar as such Article is consistent with this Agreement. The Contractor shall be reimbursed for Changes in the Work on the basis of Cost of the Work as defined in Article 9.

8.2   The Contractor's Fee for Changes in the Work shall be as set forth in Paragraph 7.2, or in the absence of specific provisions therein, shall be adjusted by negotiation on the basis of the Fee established for the original Work.

ARTICLE 9

COSTS TO BE REIMBURSED

9.1   The term Cost of the Work shall mean costs necessarily incurred in the proper performance of the Work and paid by the Contractor. Such costs shall be at rates not higher than the standard paid in the locality of the Work except with prior consent of the Owner, and shall include the items set forth below in this Article 9.

9.1.1   Wages paid for labor in the direct employ of the Contractor in the performance of the Work under applicable collective bargaining agreements, or under a salary or wage schedule agreed upon by the Owner and Contractor, and including such welfare or other benefits, if any, as may be payable with respect thereto.

9.1.2   Salaries of Contractor's employees when stationed at the field office, in whatever capacity employed. Employees engaged, at shops or on the road, in expediting the production or transportation of materials or equipment, shall be considered as stationed at the field office and their salaries paid for that portion of their time spent on this Work.

9.1.3   Cost of contributions, assessments or taxes for such items as unemployment compensation and social security, insofar as such cost is based on wages, salaries, or other remuneration paid to employees of the Contractor and included in the Cost of the Work under Subparagraphs 9.1.1 and 9.1.2.

9.1.4   The proportion of reasonable transportation, traveling and hotel expenses of the Contractor or of his officers or employees incurred in discharge of duties connected with the work.

9.1.5   Cost of all materials, supplies and equipment incorporated in the Work, including costs of transportation thereof.

9.1.6   Payments made by the Contractor to Subcontractors for Work performed pursuant to subcontracts under this Agreement.

9.1.7   Cost, including transportation and maintenance, of all materials, supplies, equipment, temporary facilities and hand tools not owned by the workmen, which are consumed in the performance of the Work, and cost less salvage value on such items used but not consumed which remain the property of the Contractor.

9.1.8   Rental charges of all necessary machinery and equipment, exclusive of hand tools, used at the site of the Work, whether rented from the Contractor or others, including installation, minor repairs and replacements, dismantling, removal, transportation and delivery costs thereof, at rental charges consistent with those prevailing in the area.

9.1.9   Cost of premiums for all bonds and insurance which the Contractor is required by the Contract Documents to purchase and maintain.

9.1.10   Sales, use or similar taxes related to the Work and for which the Contractor is liable imposed by any governmental authority.

9.1.11   Permit fees, royalties, damages for infringement of patents and costs of defending suits therefor, and deposits lost for causes other than the Contractor's negligence.

9.1.12   Losses and expenses, not compensated by insurance or otherwise, sustained by the Contractor in connection with the Work, provided they have resulted from causes other than the fault or neglect of the Contractor. Such losses shall include settlements made with the written consent and approval of the Owner. No such losses and expenses shall be included in the Cost of the Work for the purpose of determining the Contractor's Fee. If, however, such loss requires reconstruction and the Contractor is placed in charge thereof, he shall be paid for his services a Fee proportionate to that stated in Paragraph 7.1.

9.1.13   Minor expenses such as telegrams, long distance telephone calls, telephone service at the site, expressage, and similar petty cash items in connection with the Work.

9.1.14   Cost of removal of all debris.

9.1.15   Costs incurred due to an emergency affecting the safety of persons and property.

9.1.16   Other costs incurred in the performance of the Work if and to the extent approved in advance in writing by the Owner.

ARTICLE 10

COSTS NOT TO BE REIMBURSED

10.1   The term Cost of the Work shall not include any of the items set forth below in this Article 10.

10.1.1   Salaries or other compensation of the Contractor's officers, executives, general managers, estimators, auditors, accountants, purchasing and contracting agents and other employees at the Contractor's principal office and branch offices, except employees of the Contractor when engaged at shops or on the road in expediting the production or transportation of materials or equipment for the Work.

10.1.2   Expenses of the Contractor's Principal and Branch Offices other than the Field Office.

10.1.3   Any part of the Contractor's capital expenses, including interest on the Contractor's capital employed for the Work.

10.1.4   Overhead or general expenses of any kind, except as may be expressly included in Article 9.

10.1.5   Costs due to the negligence of the Contractor, any Subcontractor, anyone directly or indirectly employed by any of them, or for whose acts any of them may be liable, including but not limited to the correction of defective Work, disposal of materials and equipment wrongly supplied, or making good any damage to property.

10.1.6   The cost of any item not specifically and expressly included in the items described in Article 9.

10.1.7   Costs in excess of the Guaranteed Maximum Cost, if any, as set forth in Article 6 and adjusted pursuant to Article 8.

ARTICLE 11

DISCOUNTS, REBATES AND REFUNDS

All cash discounts shall accrue to the Contractor unless the Owner deposits funds with the Contractor with which to make payments, in which case the cash discounts shall accrue to the Owner. All trade discounts, rebates and refunds, and all returns from sale of surplus materials and equipment shall accrue to the Owner, and the Contractor shall make provisions so that they can be secured.

(Here insert any provisions relating to deposits by the Owner to permit the Contractor to obtain cash discounts.)

ARTICLE 12

SUBCONTRACTS

12.1   All portions of the Work that the Contractor's organization has not been accustomed to perform shall be performed under subcontracts. The Contractor shall request bids from subcontractors and shall deliver such bids to the Architect. The Architect will then determine, with the advice of the Contractor and subject to the approval of the Owner, which bids will be accepted.

12.2   All Subcontracts shall conform to the requirements of Paragraph 5.3 of the General Conditions. Subcontracts awarded on the basis of the cost of such work plus a fee shall also be subject to the provisions of this Agreement insofar as applicable.

ARTICLE 13

ACCOUNTING RECORDS

The Contractor shall check all materials, equipment and labor entering into the Work and shall keep such full and detailed accounts as may be necessary for proper financial management under this Agreement, and the system shall be satisfactory to the Owner. The Owner shall be afforded access to all the Contractor's records, books, correspondence, instructions, drawings, receipts, vouchers, memoranda and similar data relating to this Contract, and the Contractor shall preserve all such records for a period of three years after the final payment.

ARTICLE 14

APPLICATIONS FOR PAYMENT

The Contractor shall, at least ten days before each progress payment falls due, deliver to the Architect a statement, sworn to if required, showing in complete detail all moneys paid out or costs incurred by him on account of the Cost of the Work during the previous month for which he is to be reimbursed under Article 6 and the amount of the Contractor's Fee due as provided in Article 7, together with payrolls for all labor and all receipted bills for which payment has been received.

ARTICLE 15

PAYMENTS TO THE CONTRACTOR

15.1   The Architect will review the Contractor's statement of moneys due as provided in Article 14 and will promptly issue a Certificate for Payment to the Owner for such amount as he approves, which Certificate shall be payable on or about the   day of the month.

15.2   Final payment, constituting the unpaid balance of the Cost of the Work and of the Contractor's Fee, shall be paid by the Onwer to the Contractor when the Work has been completed, the Contract fully performed and a final Certificate for Payment has been issued by the Architect. Final payment shall be due   days after the date of issuance of the final Certificate for Payment.

ARTICLE 16

TERMINATION OF THE CONTRACT

16.1   The Contract may be terminated by the Contractor as provided in Article 14 of the General Conditions.

16.2   If the Owner terminates the Contract as provided in Article 14 of the General Conditions, he shall reimburse the Contractor for any unpaid Cost of the Work due him under Article 6, plus (1) the unpaid balance of the Fee computed upon the Cost of the Work to the date of termination at the rate of the percentage named in Article 7, or (2) if the Contractor's Fee be stated as a fixed sum, such an amount as will increase the payments on account of his Fee to a sum which bears the same ratio to the said fixed sum as the Cost of the Work at the time of termination bears to the adjusted Guaranteed Maximum Cost, if any, otherwise to a reasonable estimated Cost of the Work when completed. The Owner shall also pay to the Contractor fair compensation, either by purchase or rental at the election of the Owner, for any equipment retained. In case of such termination of the Contract the Owner shall further assume and become liable for obligations, commitments and unsettled claims that the Contractor has previously undertaken or incurred in good faith in connection with said Work. The Contractor shall, as a condition of receiving the payments referred to in this Article 16, execute and deliver all such papers and take all such steps, including the legal assignment of his contractual rights, as the Owner may require for the purpose of fully vesting in him the rights and benefits of the Contractor under such obligations or commitments.

ARTICLE 17

MISCELLANEOUS PROVISIONS

17.1   Terms used in this Agreement which are defined in the Conditions of the Contract shall have the meanings designated in those Conditions.

17.2   The Contract Documents, which constitute the entire agreement between the Owner and the Contractor, are listed in Article 1 and, except for Modifications issued after execution of this Agreement, are enumerated as follows:

(List below the Agreement, Conditions of the Contract, (General, Supplementary, other Conditions), Drawings, Specifications, Addenda and accepted Alternates, showing page or sheet numbers in all cases and dates where applicable.)

This Agreement executed the day and year first written above.

OWNER _____ CONTRACTOR _____

Appendix

A. The Standards of Professional Practice, American Institute of Architects, AIA Document J330.

B. Code of Ethical Conduct for Members of the Associated General Contractors of America.

C. Construction Industry Arbitration Rules, American Arbitration Association.

D. Policy Statement on Building Product Development and Uses, American Institute of Architects.

THE AMERICAN INSTITUTE OF ARCHITECTS

AIA Document J 330 Revised May 1967

The Standards of Professional Practice

*The following Provisions of the Bylaws
of The Institute form the basis for all disciplinary actions taken
under the Standards of Professional Practice:*

CHAPTER 14, ARTICLE 1, SECTION 1 (c) — Any deviation by a corporate member from any of the Standards of Professional Practice of The Institute or from any of the Rules of The Board supplemental thereto, or any action by him that is detrimental to the best interests of the profession and The Institute shall be deemed to be unprofessional conduct on his part, and ipso facto he shall be subject to discipline by The Institute.

PREFACE

0.1   The profession of architecture calls for men of integrity, culture, acumen, creative ability and skill. The services of an architect may include any services appropriate to the development of man's physical environment, provided that the architect maintains his professional integrity and that his services further the ultimate goal of creating an environment of orderliness and beauty. The architect's motives, abilities and conduct always must be such as to command respect and confidence.

An architect should seek opportunities to be of constructive service in civic affairs, and to advance the safety, health, beauty and well-being of his community in which he resides or practices. As an architect, he must recognize that he has moral obligations to society beyond the requirements of law or business practices. He is engaged in a profession which carries important responsibilities to the public and, therefore, in fulfilling the needs of his client, the architect must consider the public interest and the well-being of society.

0.2   An architect's honesty of purpose must be above suspicion; he renders professional services to his client and acts as his client's agent and adviser. His advice to his client must be sound and unprejudiced, as he is charged with the exercise of impartial judgment in interpreting contract documents.

0.3   Every architect should constribute generously of his time and talents to foster justice, courtesy, and sincerity within his profession. He administers and coordinates the efforts of his professional associates, subordinates, and consultants, and his acts must be prudent and knowledgeable.

0.4   Building contractors and their related crafts and skills are obligated to follow the architect's directions as expressed in the contract documents; these directions must be clear, concise, and fair.

OBLIGATIONS

1.   To the Public

1.1   An architect may offer his services to anyone on the generally accepted basis of commission, fee, salary, or royalty, as agent, consultant, adviser, or assistant, provided that he strictly maintains his professional integrity.

1.2   An architect shall perform his professional services with competence, and shall properly serve the interests of his client and the public.

1.3   An architect shall not engage in building contracting.

1.4   An architect shall not use paid advertising or indulge in self-laudatory, exaggerated, misleading or false publicity, nor shall he publicly endorse products or permit the use of his name to imply endorsement.

1.5   An architect shall not solicit, nor permit others to solicit in his name, advertisements or other support toward the cost of any publication presenting his work.

1.6   An architect shall conform to the registration laws governing the practice of architecture in any state in which he practices, and shall observe the customs and standards established by the local professional body of architects.

2.   To the Client

2.1   An architect's relation to his client is based upon the concept of agency. Before undertaking any commission he shall determine with his client the scope of the project, the nature and extent of the services he will perform and his compensation for them, and shall provide communication thereof in writing. In performing his services he shall maintain an understanding with his client regarding the project, its developing solutions and its estimated probable costs. Where a fixed limit of cost is established in advance of design, the architect must determine the character of design construction so as to meet as nearly as feasible the cost limit established. He shall keep his client informed with competent estimates of probable costs. He shall not guarantee the final cost, which will be determined not only by the architect's solution of the owner's requirements, but by the fluctuating conditions of the competitive construction market.

2.2   An architect shall guard the interest of his client and the rights of those whose contracts the architect administers. An architect should give every reasonable aid toward a complete understanding of those contracts in order that mistakes may be avoided.

2.3   An architect's communications, whether oral, written, or graphic, should be definite and clear.

2.4   An architect shall not have financial or personal interests which might tend to compromise his obligation to his client.

2.5   An architect shall not accept any compensation for his professional services from anyone other than his client or employer.

2.6   An architect shall base his compensation on the value of the services he agrees to render. He shall neither offer nor agree to perform his services for a compensation that will tend to jeopardize the adequacy or professional quality of those services, or the judgment, care and diligence necessary properly to discharge his responsibilities to his client and the public.

3. To the Profession

3.1   A member shall support the interests, objectives and Standards of Professional Practice of The American Institute of Architects.

3.2   An architect shall not act in a manner detrimental to the best interests of the profession.

3.3   An architect shall not knowingly injure or attempt to injure falsely or maliciously the professional reputation, prospects, or practice of another architect.

3.4   An architect shall not attempt to supplant another architect after definite steps have been taken by a client toward the latter's employment. He shall not offer to undertake a commission for which he knows another architect has been employed, nor shall he undertake such a commission until he has notified such other architect of the fact in writing, and has been advised by the owner that employment of that architect has been terminated.

3.5   An architect shall not enter into competitive bidding against another architect on the basis of compensation. He shall not use donation or misleading information on cost as a device for obtaining a competitive advantage.

3.6   An architect shall not offer his services in a competition except as provided in the Competition Code of The American Institute of Architects.

3.7   An architect shall not engage a commission agent to solicit work in his behalf.

3.8   An architect shall not call upon a contractor to provide work to remedy omissions or errors in the contract documents without proper compensation to the contractor.

3.9   (Note: This paragraph was deleted by the action of the 1967 Convention.)

3.10   An architect shall not be, nor continue to be, a member of a firm or organization or an employee of an individual, firm or organization which practices in a manner inconsistent with these Standards of Professional Practice.

3.11   Dissemination by an architect, or by any component of The Institute, of information concerning judiciary procedures and penalties, beyond the information published or authorized by The Board or its delegated authority, shall be considered to be detrimental to the best interests of the architectural professional.

4. To Related Professionals

4.1   An architect should provide his professional employees with a desirable working environment and compensate them fairly.

4.2   An architect should contribute to the interchange of technical information and experience between architects, the design profession, and the building industry.

4.3   An architect should respect the interests of consultants and associated professionals in a manner consistent with the applicable provisions of these Standards of Practice.

4.4   An architect should recognize the contribution and the professional stature of the related professionals and should collaborate with them in order to create an optimum physical environment.

4.5   An architect should promote interest in the design professions and facilitate the professional development of those in training. He should encourage a continuing education, for himself and others, in the functions, duties, and responsibilities of the design professions, as well as the technical advancement of the art and science of environmental design.

PROMULGATION

5.1   These Standards of Professional Practice are promulgated to promote the highest ethical standards for the profession of architecture. Thus the enumeration of particular duties in the Standards should not be construed as a denial of others, equally imperative, though not specifically mentioned. Furthermore, the placement of statements of obligation under any category above shall not be construed as limiting the applicability of such statement to the group named, since some obligations have broad application, and have been positioned as they are only as a matter of convenience and emphasis. The primary purpose of disciplinary action under these Standards of Professional Practice is to protect the public and the profession.

5.2   Since adherence to the principles herein enumerated is the obligation of every member of The American Institute of Architects, any deviation therefrom shall be subject to discipline in proportion to its seriousness.

5.3   The Board of Directors of The American Institute of Architects, or its delegated authority, shall have sole power of interpreting these Standards of Professional Practice and its decisions shall be final subject to the provisions of the Bylaws.

NOTE: This 1967 edition of the Standards of Professional Practice differs from the 1965 edition in the wording of Paragraphs 3.9 and 3.10.

CODE OF ETHICAL CONDUCT

THE ASSOCIATED GENERAL CONTRACTORS OF AMERICA

NATIONAL HEADQUARTERS, 1957 E STREET, N.W., WASHINGTON, D. C. 20006

CODE OF ETHICAL CONDUCT

*Adopted by The Associated General Contractors of America,
January 1925, with a preamble by its Committee on Ethics*

PREAMBLE

During its comparatively short life, The Associated General Contractors of America has achieved among the older professional and industrial organizations a position of confidence and respect, which it earnestly desires to maintain. This position has been attained principally by persevering effort to promote scientific construction methods, to rid the industry of improper practices and to eliminate waste in its operations. Scrupulous avoidance of any action which might not prove compatible with the public interest or with the rightful interest of other industries together with the practice of opening all Conventions and Directors Meetings to the public have helped to demonstrate that its purposes are sound and worthy.

As an Association, we recognize and seek those benefits which accrue from compliance with high ethical standards.

FUNCTION OF ETHICS

The function of ethical standards for a society such as this is essentially twofold; first, to establish principles of business conduct to be observed by the members in their relations with each other, and second, to establish principles to be observed in their transactions with those who utilize their services. In either sense they represent the minimum requirements for fair competition and honorable dealing.

The philosophy of ethical conduct is not new, nor is it unfamiliar to those men who have grasped the purpose of this Association. It is not necessary therefore, to dwell upon our obligation to declare and follow what years of commercial experience have indicated as sound ethical practice. There are, however, certain practical considerations involved in this matter, which every contractor, and in fact every business man, should recognize.

PRACTICAL CONSIDERATIONS

The business structure of America founded upon private initiative, has grown so complex that successful government of business conduct by legislation is not only undesirable but practically impossible. Industries and individuals, in order to retain the benefits arising from this free initiative, and to avoid the hampering burdens of regulatory legislation, must themselves accept responsibility for fair and intelligent self-government. Such government does not require a profound knowledge of law and legal procedure, but merely the courage to abide by self-imposed restraints and principles which the intelligent fair-minded majority of any industry clearly understand. Especially in construction, which affects so fundamentally the entire country, is this courage needed.

The Associated General Contractors of America realizes that the vital bearing of the construction industry upon the well-being, comfort, and safety of the entire public injects into the contractor's function an element of professional responsibility founded upon honor and trust. This responsibility requires, among other things, that we seek to improve construction methods, management and service, to eliminate uneconomical and improper practices, and to build responsibility throughout our industry. It surely cannot mean less than the establishment of construction service which will give to the investing public an assurance of skill and faithful performance.

ACCEPTABLE PRINCIPLES

There are a number of principles to which the members of this Association can subscribe, but which have not yet been clearly formulated. We can accept without reservation the principle recognized by fair-minded men in every industry that contracts whether written or oral should be carried out according to their bona fide intent, and that disagreement concerning their intent should be settled, if possible, by impartial arbitration.

The attitude of contractors toward their workmen is an element of ethics or fair dealing, and we recognize the vital relation of the welfare and success of contractors to that of their workmen. They should, in their dealings with labor, be guided by principles of justice and fairness, recognizing in this relation high standards of conduct and craftsmanship.

The Associated General Contractors of America limits its membership to contractors who qualify as to SKILL, INTEGRITY, and RESPONSIBILITY, and whose dealings are honorable and ethical; and it is perhaps superfluous to suggest that no individual possessing these qualifications would attempt to injure the professional reputation, prospects or business of a competitor.

NEED OF INTERPRETATION

Thus in a measure may be outlined some of the fundamental conceptions of construction ethics which we realize must be clearly expressed and generally accepted if this industry is to maintain its rightful place in the industrial world. Mere acceptance of the principles, however, no matter how sincerely accepted does not assure practical application in the daily transactions of business. Individual minds make individual interpretations and, unless there is some elucidation of principle the many minds do not meet and the principle fails in application. Otherwise we should need only the Golden Rule to govern all human activities.

In developing ethical standards, choice lies somewhere between the shortest possible statement and a complete, explicit set of rules attempting to cover all eventualities of the industry. Each of these conceptions has found favor within the membership of the Association. Approval has already been given to the Golden Rule, and certainly all sound ethical codes must rest upon that inspired foundation, but to go no further would leave this very practical industry at the mercy of individual interpretation and consequently without any standard rules of conduct.

Yet it seems undesirable and well nigh impossible to elaborate upon specific points and to develop a comprehensive structure which would completely provide for all questions and phases of the contracting business. This must develop slowly out of experience, if at all. Consequently the Committee on Ethics rather briefly suggests the broad principles upon which we stand, and presents herewith some essential rules upon which we may all approximately agree.

No code will be final except in its fundamental principles. Our rules of practice will inevitably be modified or added to as experience develops. As ideas on conduct crystallize, various rules which now are merely suggestive may become mandatory, or may be abandoned. Therefore, while these rules of ethical practice are the result of careful thought and study, they are not considered as either complete or perfect, but are offered as a starting point or frame work which appears fairly adequate today and upon which a safe ethical structure may in time be developed.

RULES OF ETHICAL PRACTICE

The working principles by which members of The Associated General Contractors are to be governed in their relations with client owners and the public, with other agencies of construction, and with members of their own profession are as follows:

1.   OWNERS AND THE PUBLIC

Fair and bona fide competition is a fundamental service of our industry to which clients and owners are entitled. Any act or method in restriction thereof is a breach of faith toward this Association and a betrayal of its principles.

But the competition cannot serve its legitimate purpose unless it operates under conditions alike fair to owner and contractor.

Observance of ethical conduct toward the contractor by those who utilize his competitive bidding will be encouraged in proportion as he himself abides by the ethics of fair competition. Only when he respects the code of this Association can be reasonably ask others to respect it.

Ethical conduct with respect to competitive bidding is defined in the following paragraphs:

1. Competitive bids preferably should be submitted only when a definite time and place for the opening of all proposals has been fixed, at which all bidders or their representatives are permitted to be present.

2. The contractor's professional knowledge is the result of his training and experience and if he is called upon for preliminary estimates or appraisals it is proper that he should be paid in the same manner that engineers and architects are paid for similar service.

3. Bidders should neither seek nor accept information concerning a competitor's bid prior to the opening, nor by any method suppress free competition. It is equally improper for the owners to use bids in an effort to induce any contractor to lower his figures.

4. On private work, if all competitive bids are rejected, new bids should not be submitted within 60 days unless warranted by a substantial change either in the work to be performed, or the market, or other basic conditions affecting cost.

5. The amount of a bid should not be altered after the opening except when substantial change is made in the work, or when further bidding on alternate items is requested. In event of such change or further bidding, the contractor should bid only on the items specified and should increase or decrease the amount of his bid only in proportion to the cost of the change or alternate involved. Any reduction of a bid disproportionate to such cost or the submission of any alternate which in effect produces such a disproportionate reduction, constitutes unfair competition. This shall not be construed as prohibiting the low bidder from decreasing his bid after he has ascertained that he is the low bidder.

6. Contractors should cooperate in advising architects, engineers and owners with respect to the relative costs of various alternates while plans are being prepared and thus seek to reduce the number of alternates to a nominal maximum.

7. When bids are solicited and received by an owner on a lump sum basis, no competitor other than the low bidder should solicit the work on a percentage basis, or any other form of cost-plus contract, provided however, that any competitor shall have the right to accept the work at his bid price or on a percentage basis if tendered him

without guaranteed maximum cost or at a guaranteed maximum cost not less than his original bid.

2. ENGINEERING AND ARCHITECTURAL PROFESSIONS

Local and national cooperation in matters of mutual concern should be the basic policy of members of this Association in their relations with the engineering and architectural professions; the purpose of this cooperation being to establish a clear conception of respective functions and responsibilities, to guard against uneconomical or improper practices, and to carry out constructive measures within the industry.

Ethical conduct toward architects and engineers demands the following:

1. Support should be given to all efforts of these professions to maintain and extend high standards of conduct.
2. Contractors should give full credit to the value of the services rendered by the architect and engineer and neither undermine nor disparage their functions and usefulness.

3. SUB-CONTRACTORS AND THOSE WHO SUPPLY MATERIALS

The operations of the contractor are made possible through the functioning of those agencies which furnish him with service or products, and in contracting with them he is rightfully obligated by the same principles of honor and fair dealing that he desires should govern the actions toward himself of architects, engineers and client owners.

Ethical conduct with respect to sub-contractors and those who supply materials requires that:

1. Proposals should not be invited from anyone who is known to be unqualified to perform the proposed work or to render the proper service.
2. The figures of one competitor shall not be made known to another before the award of the subcontract, nor should they be used by the contractor to secure a lower proposal from another bidder.
3. The contract should preferably be awarded to the lowest bidder if he is qualified to perform the contract, but if the award is made to another bidder, it should be at the amount of the latter's bid.
4. In no case should the low bidder be led to believe that a lower bid than his has been received.
5. When the contractor has been paid by a client owner for work or material, he should make payment promptly, and in just proportion, to subcontractors and others.

4. OPERATING WITHIN THE JURISDICTION OF A CHAPTER OR BRANCH

The conduct of any member who traditionally contracts with the building trades unions when operating within the jurisdiction of any Chapter or Branch whose members play a principal role in the local bargaining unit should comply with the following:

1. Prior to estimating work within that area, he should first contact the appropriate local chapter or Branch headquarters which should furnish him with complete information as to local conditions, prevailing scale of wages and working conditions which prevail on the proposed work.

2. If awarded work in that area he should become a member of that appropriate local Chapter or Branch to the end that he may perform his work under the conditions and scale of wages established under the jurisdiction of that local Chapter or Branch.
3. Any action which tends to undermine the integrity of the local bargaining unit such as the use of the interim short-form agreement and that section of a national agreement which requires the contractor to continue operations during strikes against, or lockouts by local contractors, should not be used.

5. DISCIPLINE

Any member who refuses to abide by this Code of Ethical Practice shall be subject to discipline as provided by Article IV-F, Section 3, of the Governing Provisions.

Note: Article 4 is printed as revised March 1968

CONSTRUCTION

INDUSTRY

ARBITRATION RULES

Effective March 8, 1966

AMERICAN INSTITUTE OF ARCHITECTS

ASSOCIATED GENERAL CONTRACTORS

CONSULTING ENGINEERS COUNCIL

**COUNCIL OF MECHANICAL SPECIALTY
CONTRACTING INDUSTRIES**

**NATIONAL SOCIETY OF
PROFESSIONAL ENGINEERS**

ADMINISTERED BY

AMERICAN ARBITRATION ASSOCIATION

140 West 51st Street   New York, N. Y. 10020

AMERICAN ARBITRATION ASSOCIATION

Arbitration is the voluntary submission of a dispute to a disinterested person or persons for final determination. And to achieve orderly, economical and expeditious arbitration, in accordance with federal and state laws, the American Arbitration Association is available to administer arbitration cases under various specialized rules.

The American Arbitration Association maintains throughout the United States a National Panel of Arbitrators consisting of experts in all trades and professions. By arranging for arbitration under the Construction Industry Arbitration Rules, parties may obtain the services of arbitrators who are familiar with the construction industry.

The American Arbitration Association shall establish and maintain as members of its National Panel of Arbitrators individuals competent to hear and determine disputes administered under the Construction Industry Arbitration Rules. The Association shall consider for appointment to the Construction Industry Panel persons recommended by the Construction Industry National Committee as qualified to serve by virtue of their experience in the construction field.

The Association does not act as arbitrator. Its functions is to administer arbitrations in accordance with the agreement of the parties and to maintain Panels from which arbitrators may be chosen by parties. Once designated, the arbitrator decides the issues and his award is final and binding.

When an agreement to arbitrate is written into a construction contract, it may expedite peaceful settlement without the necessity of going to arbitration at all. Thus, the arbitration clause is a form of insurance against loss of good will.

TABLE OF CONTENTS

Note: Numbers at right refer to pages in American Arbitration Association publication AAA-47-10M-3-67.

CONSTRUCTION INDUSTRY ARBITRATION RULES

Section 1. AGREEMENT OF PARTIES – The parties shall be deemed to have made these Rules a part of their arbitration agreement whenever they have provided for arbitration under the Construction Industry Arbitration Rules. These Rules and any amendment thereof shall apply in the form obtaining at the time the arbitration is initiated.

Section 2. NAME OF TRIBUNAL – Any Tribunal constituted by the parties for the settlement of their dispute under these Rules shall be called the Construction Industry Arbitration Tribunal, hereinafter called the Tribunal.

Section 3. ADMINISTRATOR – When parties agree to arbitrate under these Rules, or when they provide for arbitration by the American Arbitration Association, hereinafter called AAA, and an arbitration is initiated hereunder, they thereby constitute AAA the administrator of the arbitration. The authority and duties of the administrator are prescribed in the agreement of the parties and in these Rules.

Section 4. DELEGATION OF DUTIES – The duties of the AAA under these Rules may be carried out through Tribunal Clerks, or such other officers of committees as the AAA may direct.

Section 5. NATIONAL PANEL OF ARBITRATORS – In cooperation with the Construction Industry Arbitration Committee, the AAA shall establish and maintain a National Panel of Construction Arbitrators, hereinafter called the Panel, and shall appoint an arbitrator or arbitrators therefrom as hereinafter provided. A neutral arbitrator selected by mutual choice of both parties or their appointees, or appointed by the AAA, is hereinafter called the arbitrator, whereas an arbitrator selected unilaterally by one party is hereinafter called the party-appointed arbitrator. The term Arbitrator may hereinafter be used to refer to one arbitrator or to a Tribunal of multiple arbitrators.

Section 6. OFFICE OF TRIBUNAL – The general office of a Tribunal is the headquarters of the AAA, which may, however, assign the administration of an arbitration to any of its Regional Offices.

Section 7. INITIATION UNDER AN ARBITRATION PROVISION IN A CONTRACT – Arbitration under an arbitration provision in a contract shall be initiated in the following manner:

The initiating party shall, within the time specified by the contract, if any, file with the other party a notice of his intention to arbitrate (Demand), which notice shall contain a statement setting forth the nature of the dispute, the amount involved, if any, and the remedy sought; and shall file two copies of said notice with any Regional Office of the AAA, together with two copies of the arbitration provisions of the contract and the appropriate filing fee as provided in Section 47 hereunder.

The AAA shall give notice of such filing to the other party. If he so desires, the party upon whom the demand for arbitration is made may file an answering statement in duplicate with the AAA within seven days after notice from the AAA, in which event he shall simultaneously send a copy of his answer to the other party. If a monetary claim is made in the answer the appropriate administrative fee provided in the Fee Schedule shall be forwarded to the AAA with the answer. If no answer is filed within the stated time, it will be treated as a denial of the claim. Failure to file an answer shall not operate to delay the arbitration.

Section 8. CHANGE OF CLAIM – After filing of the claim, if either party desires to make any new or different claim, such claim shall be made in writing and filed with the

AAA, and a copy thereof shall be mailed to the other party who shall have a period of seven days from the date of such mailing within which to file an answer with the AAA. However, after the Arbitrator is appointed no new or different claim may be submitted to him except with his consent.

Section 9. INITIATION UNDER A SUBMISSION – Parties to any existing dispute may commence an arbitration under these Rules by filing at any Regional Office two (2) copies of a written agreement to arbitrate under these Rules (Submission), signed by the parties. It shall contain a statement of the matter in dispute, the amount of money involved, if any, and the remedy sought, together with the appropriate filing fee as provided in the Fee Schedule.

Section 10. FIXING OF LOCALE – The parties may mutually agree on the locale where the arbitration is to be held. If any party requests that the hearing be held in a specific locale and the other party files no objection thereto within seven days after notice of the request is mailed to such party, the locale shall be the one requested. If a party objects to the locale requested by the other party, the AAA shall have power to determine the locale and its decision shall be final and binding.

Section 11. QUALIFICATIONS OF ARBITRATOR – No person shall serve as an Arbitrator in any arbitration if he has any financial or personal interest in the result of the arbitration, unless the parties, in writing, waive such disqualification.

Section 12. APPOINTMENT FROM PANEL – If the parties have not appointed an Arbitrator and have not provided any other method of appointment, the Arbitrator shall be appointed in the following manner: Immediately after the filing of the Demand or Submission, the AAA shall submit simultaneously to each party to the dispute an identical list of names of persons chosen from the Panel. Each party to the dispute shall have seven days from the mailing date in which to cross off any name to which he objects, number the remaining names indicating the order of his preference, and return the list to the AAA. If a party does not return the list within the time specified, all persons named therein shall be deemed acceptable. From among the persons who have been approved on both lists, and in accordance with the designated order of mutual preference, the AAA shall invite the acceptance of an Arbitrator to serve. If the parties fail to agree upon any of the persons named, or if acceptable Arbitrators are unable to act, or if for any other reason the appointment cannot be made from the submitted lists, the AAA shall have the power to make the appointment from other members of the Panel without the submission of any additional lists.

Section 13. DIRECT APPOINTMENT BY PARTIES – If the agreement of the parties names an Arbitrator or specifies a method of appointing an Arbitrator, that designation or method shall be followed. The notice of appointment, with name and address of such Arbitrator, shall be filed with the AAA by the appointing party. Upon the request of any such appointing party, the AAA shall submit a list of members from the Panel from which the party may, if he so desires, make the appointment.

If the agreement specifies a period of time within which an Arbitrator shall be appointed, and any party fails to make such appointment within that period, the AAA shall make the appointment.

If no period of time is specified in the agreement, the AAA shall notify the parties to make the appointment and if within seven days after mailing of such notice such Arbitrator has not been so appointed, the AAA shall make the appointment.

Section 14. APPOINTMENT OF ARBITRATOR BY PARTY-APPOINTED ARBITRATORS – If the parties have appointed their party-appointed Arbitrators or if either or both of them have been appointed as provided in Section 13, and have authorized such Arbitrators to appoint an Arbitrator within a specified time and no appointment is made within such time or any agreed extension thereof, the AAA shall appoint an Arbitrator who shall act as Chairman.

If no period of time is specified for appointment of the third Arbitrator and the party-appointed Arbitrators do not make the appointment within seven days from the date of the appointment of the last party-appointed Arbitrator, the AAA shall appoint the Arbitrator who shall act as Chairman.

If the parties have agreed that their party-appointed Arbitrators shall appoint the Arbitrator from the Panel, the AAA shall furnish to the party-appointed Arbitrators, in the manner prescribed in Section 12, a list selected from the Panel, and the appointment of the Arbitrator shall be made as prescribed in such Section.

Section 15. NATIONALITY OF ARBITRATOR IN INTERNATIONAL ARBITRATION — If one of the parties is a national or resident of a country other than the United States, the Arbitrator shall, upon the request of either party, be appointed from among the nationals of a country other than that of any of the parties.

Section 16. NUMBER OF ARBITRATORS — If the arbitration agreement does not specify or the parties are unable to agree as to the number of Arbitrators, the dispute shall be heard and determined by three Arbitrators, unless the AAA, in its discretion, directs that a single Arbitrator or a greater number of Arbitrators be appointed.

Section 17. NOTICE TO ARBITRATOR OF HIS APPOINTMENT — Notice of the appointment of the Arbitrator, whether mutually appointed by the parties or by the AAA, shall be mailed to the Arbitrator by the AAA, together with a copy of these Rules, and the signed acceptance of the Arbitrator shall be filed prior to the opening of the first hearing.

Section 18. DISCLOSURE BY ARBITRATOR OF DISQUALIFICATION — Prior to accepting his appointment, the prospective Arbitrator shall disclose any circumstances likely to create a presumption of bias or which he believed might disqualify him as an impartial Arbitrator. Upon receipt of such information, the AAA shall immediately disclose it to the parties who, if willing to proceed under the circumstances disclosed, shall so advise the AAA in writing. If either party declines to waive the presumptive disqualification, the vacancy thus created shall be filled in accordance with the applicable provisions of these Rules.

Section 19. VACANCIES — If any Arbitrator should resign, die, withdraw, refuse, be disqualified or be unable to perform the duties of his office, the AAA shall, on proof satisfactory to it, declare the office vacant. Vacancies shall be filled in accordance with the applicable provisions of these Rules and the matter shall be reheard unless the parties shall agree otherwise.

Section 20. TIME AND PLACE — The Arbitrator shall fix the time and place for each hearing. The AAA shall mail to each party notice thereof at least five days in advance, unless the parties by mutual agreement waive such notice or modify the terms thereof.

Section 21. REPRESENTATION BY COUNSEL — Any party may be represented by counsel. A party intending to be so represented shall notify the other party and the AAA of the name and address of counsel at least three days prior to the date set for the hearing at which counsel is first to appear. When an arbitration is initiated by counsel, or where an attorney replies for the other party, such notice is deemed to have been given.

Section 22. STENOGRAPHIC RECORD — The AAA shall make the necessary arrangements for the taking of a stenographic record whenever such record is requested by a party. The requesting party or parties shall pay the cost of such record as provided in Section 49.

Section 23. INTERPRETER – The AAA shall make the necessary arrangements for the services of an interpreter upon the request of one or both parties, who shall assume the cost of such service.

Section 24. ATTENDANCE AT HEARINGS – Persons having a direct interest in the arbitration are entitled to attend hearings. The Arbitrator shall otherwise have the power to require the retirement of any witness or witnesses during the testimony of other witnesses. It shall be discretionary with the Arbitrator to determine the propriety of the attendance of any other persons.

Section 25. ADJOURNMENTS – The Arbitrator may take adjournments upon the request of a party or upon his own initiative and shall take such adjournment when all of the parties agree thereto.

Section 26. OATHS – Before proceeding with the first hearing or with the examination of the file, each Arbitrator may take an oath of office, and if required by law, shall do so. The Arbitrator may, in his discretion, require witnesses to testify under oath administered by any duly qualified person or, if required by law or demanded by either party, shall do so.

Section 27. MAJORITY DECISION – Whenever there is more than one Arbitrator, all decisions of the Arbitrators must be by at least a majority. The award must also be made by at least a majority unless the concurrence of all is expressly required by the arbitration agreement or by law.

Section 28. ORDER OF PROCEEDINGS – A hearing shall be opened by the filing of the oath of the Arbitrator, where required, and by the recording of the place, time and date of the hearing, the presence of the Arbitrator and parties, and counsel, if any, and by the receipt by the Arbitrator of the statement of the claim and answer, if any.

The Arbitrator may, at the beginning of the hearing, ask for statements clarifying the issues involved.

The complaining party shall then present his claim and proofs and his witnesses, who shall submit to questions or other examination. The defending party shall then present his defense and proofs and his witnesses, who shall submit to questions or other examination. The Arbitrator may in his discretion vary this procedure but he shall afford full and equal opportunity to the parties for the presentation of any material or relevant proofs.

Exhibits, when offered by either party, may be received in evidence by the Arbitrator.

The names and addresses of all witnesses and exhibits in order received shall be made a part of the record.

Section 29. ARBITRATION IN THE ABSENCE OF A PARTY – Unless the law provides to the contrary, the arbitration may proceed in the absence of any party, who, after due notice, fails to be present or fails to obtain an adjournment. An award shall not be made solely on the default of a party. The Arbitrator shall require the party who is present to submit such evidence as he may require for the making of an award.

Section 30. EVIDENCE – The parties may offer such evidence as they desire and shall produce such additional evidence as the Arbitrator may deem necessary to an understanding and determination of the dispute. When the Arbitrator is authorized by law to subpoena witnesses or documents, he may do so upon his own initiative or upon the request of any party. The Arbitrator shall be the judge of the admissibility of the evidence offered and conformity to legal rules of evidence shall not be necessary. All evidence shall be taken in the presence of all of the Arbitrators and all of the parties, except where any of the parties is absent in default or has waived his right to be present.

Section 31. EVIDENCE BY AFFIDAVIT AND FILING OF DOCUMENTS – The Arbitrator may receive and consider the evidence of witnesses by affidavit, but shall give it only such weight as he deems it entitled to after consideration of any objections made to its admission.

All documents not filed with the Arbitrator at the hearing, but arranged for at the hearing or subsequently by agreement of the parties, shall be filed with the AAA for transmission to the Arbitrator. All parties shall be afforded opportunity to examine such documents.

Section 32. INSPECTION OR INVESTIGATION – Whenever the arbitrator deems it necessary to make an inspection or investigation in connection with the arbitration, he shall direct the AAA to advise the parties of his intention. The Arbitrator shall set the time and the AAA shall notify the parties thereof. Any party who so desires may be present at such inspection or investigation. In the event that one or both parties are not present at the inspection or investigation, the Arbitrator shall make a verbal or written report to the parties and afford them an opportunity to comment.

Section 33. CONSERVATION OF PROPERTY – The Arbitrator may issue such orders as may be deemed necessary to safeguard the property which is the subject matter of the arbitration without prejudice to the rights of the parties or to the final determination of the dispute.

Section 34. CLOSING OF HEARINGS – The Arbitrator shall specifically inquire of the parties whether they have any further proofs to offer or witnesses to be heard. Upon receiving negative replies, the Arbitrator shall declare the hearings closed and a minute thereof shall be recorded. If briefs are to be filed, the hearings shall be declared closed as of the final date set by the Arbitrator for the receipt of briefs. If documents are to be filed as provided for in Section 31 and the date set for their receipt is later than that set for the receipt of briefs, the later date shall be the date of closing the hearing. The time limit within which the Arbitrator is required to make his award shall commence to run, in the absence of other agreements by the parties, upon the closing of the hearings.

Section 35. REOPENING OF HEARINGS – The hearings may be reopened by the Arbitrator on his own motion, or upon application of a party at any time before the award is made. If the reopening of the hearing would prevent the making of the award within the specific time agreed upon by the parties in the contract out of which the controversy has arisen, the matter may not be reopened, unless the parties agree upon the extension of such time limit. When no specific date is fixed in the contract, the Arbitrator may reopen the hearings, and the Arbitrator shall have thirty days from the closing of the reopened hearings within which to make an award.

Section 36. WAIVER OF ORAL HEARING – The parties may provide, by written agreement, for the waiver of oral hearings. If the parties are unable to agree as to the procedure, the AAA shall specify a fair and equitable procedure.

Section 37. WAIVER OF RULES – Any party who proceeds with the arbitration after knowledge that any provision or requirement of these Rules has not been complied with and who fails to state his objection thereto in writing, shall be deemed to have waived his right to object.

Section 38. EXTENSIONS OF TIME – The parties may modify any period of time by mutual agreement. The AAA for good cause may extend any period of time established by these Rules, except the time for making the award. The AAA shall notify the parties of any such extension of time and its reason therefor.

Section 39. COMMUNICATION WITH ARBITRATOR AND SERVING OF NO-TICES — There shall be no communication between the parties and an Arbitrator other than at oral hearings. Any other oral or written communications from the parties to the Arbitrator shall be directed to the AAA for transmittal to the Arbitrator.

Each party to an agreement which provides for arbitration under these Rules shall be deemed to have consented that any papers, notices or process necessary or proper for the initiation or continuation of an arbitration under these Rules and for any court action in connection therewith or for the entry of judgment on any award made thereunder may be served upon such party by mail addressed to such party or his attorney at his last known address or by personal service, within or without the state wherein the arbitration is to be held (whether such party be within or without the United States of America), provided that reasonable opportunity to be heard with regard thereto has been granted such party.

Section 40. TIME OF AWARD — The award shall be made promptly by the Arbitrator and, unless otherwise agreed by the parties, or specified by law, not later than thirty days from the date of closing the hearings, or if oral hearings have been waived, from the date of transmitting the final statements and proofs to the Arbitrator.

Section 41. FORM OF AWARD — The award shall be in writing and shall be signed either by the sole Arbitrator or by at least a majority if there be more than one. It shall be executed in the manner required by law.

Section 42. SCOPE OF AWARD — The Arbitrator may grant any remedy or relief which he deems just and equitable and within the terms of the agreement of the parties. The Arbitrator, in his award, shall assess arbitration fees and expenses as provided in §§47 and 49 equally or in favor of any party and, in the event any administrative fees or expenses are due the AAA, in favor of the AAA.

Section 43. AWARD UPON SETTLEMENT — If the parties settle their dispute during the course of the arbitration, the Arbitrator, upon their request, may set forth the terms of the agreed settlement in an award.

Section 44. DELIVERY OF AWARD TO PARTIES — Parties shall accept as legal delivery of the award the placing of the award or a true copy thereof in the mail by the AAA, addressed to such party at his last known address or to his attorney, or personal service of the award, or the filing of the award in any manner which may be prescribed by law.

Section 45. RELEASE OF DOCUMENTS FOR JUDICIAL PROCEEDINGS — The AAA shall, upon the written request of a party, furnish to such party, at his expense, certified facsimiles of any papers in the AAA's possession that may be required in judicial proceedings relating to the arbitration.

Section 46. APPLICATIONS TO COURT — No judicial proceedings by a party relating to the subject matter of the arbitration shall be deemed a waiver of the party's right to arbitrate.

The AAA is not a necessary party in judicial proceedings relating to the arbitration.

Section 47. ADMINISTRATIVE FEES — As a nonprofit organization, the AAA shall prescribe an administrative fee schedule and a refund schedule to compensate it for the cost of providing administrative services. The schedule in effect at the time of filing or the time of refund shall be applicable.

The administrative fees shall be advanced by the initiating party or parties in accordance with the administrative fee schedule, subject to final apportionment by the Arbitrator in his award.

When a matter is withdrawn or settled, the refund shall be made in accordance with the refund schedule.

The AAA, in the event of extreme hardship on the part of any party, may defer or reduce the administrative fee.

Section 48. FEE WHEN ORAL HEARINGS ARE WAIVED — Where all Oral Hearings are waived under Section 36 the Administrative Fee Schedule shall apply.

Section 49. EXPENSES — The expenses of witnesses for either side shall be paid by the party producing such witnesses.

The cost of the stenographic record, if any is made, and all transcripts thereof, shall be prorated equally between the parties ordering copies unless they shall otherwise agree and shall be paid for by the responsible parties directly to the reporting agency.

All other expenses of the arbitration, including required traveling and other expenses of the Arbitrator and of AAA representatives, and the expenses of any witness or the cost of any proofs produced at the direct request of the Arbitrator, shall be borne equally by the parties, unless they agree otherwise, or unless the Arbitrator in his award assesses such expenses or any part thereof against any specified party or parties.

Section 50. ARBITRATOR'S FEE — Unless the parties agree to the terms of compensation, members of the National Panel of Construction Arbitrators will serve without compensation for the first two days of service. Thereafter, unless the parties agree as to the terms of compensation of Arbitrators, compensation shall be based upon the amount of service involved and the number of hearings. An appropriate daily rate and other arrangements shall be discussed with the Arbitrators and submitted to the parties prior to the appointment of the Arbitrators.

Any arrangements for the compensation of an Arbitrator shall be made through the AAA and not directly by him with the parties. The terms of compensation of Arbitrators on a Tribunal shall be identical.

Section 51. DEPOSITS — The AAA may require the parties to deposit in advance such sums of money as it deems necessary to defray the expense of the arbitration, including the Arbitrator's fee if any, and shall render an accounting to the parties and return any unexpended balance.

Section 52. INTERPRETATION AND APPLICATION OF RULES — The Arbitrator shall interpret and apply these Rules insofar as they relate to his powers and duties. When there is more than one Arbitrator and a difference arises among them concerning the meaning or application of any such Rules, it shall be decided by a majority vote. If that is unobtainable, either an Arbitrator or a party may refer the question to the AAA for final decision. All other Rules shall be interpreted and applied by the AAA.

THE AMERICAN INSTITUTE OF ARCHITECTS

POLICY STATEMENT ON BUILDING PRODUCT DEVELOPMENT AND USES

PRINCIPLES

The development and introduction of new products is essential to the growth of the building industry. Equally important is a proper understanding of the characteristics, properties and appropriate uses of both new and previously existing products. Assuring the successful use and application of products as well as their related assemblies greatly minimizes the possibility, during construction and after completion, of failures which have frequently caused costly delays and grievous loss to all connected with the project.

The purpose of this policy statement is to encourage continued development of new building products and better uses for existing ones, to add to the existing technical knowledge concerning these products, and to foster better understanding between the parties involved in the building process, all being conducive to improved building design and accelerated technological progress in the entire building industry.

DEFINITION OF TERMS

The following explanatory material is for clarification of this policy statement and is not intended to be all-inclusive.

Manufacturer refers to the producer of the product, system or assembly, and includes, in addition to the prime supplier, his sales and distribution outlets and their respective representatives. The term does not include contractors independent of the organization of the original producer.

Contractor refers to independent prime contractors, subcontractors, and their employees installing specific building products.

Architect refers to the individual or organization as defined in the AIA documents, engaged in the design of construction and buildings, and includes engineers and others in the architect's employ or under his control.

OBLIGATIONS OF THE PARTIES

Manufacturer

The Manufacturer should supply the Architect with all essential data concerning his product, including pertinent information which would involve its installation, use and maintenance. In addition to the physical properties, this should take the form of chemical descriptions, laboratory and field test results, standards and ratings.

Particularly important is information on the product's compatibility and interfitting with interrelated products, as well as precautions and specific warnings on where the product should not be used based on conditions of known or anticipated failures.

The Manufacturer is expected to supply pertinent data concerning the compatibility, physical relationship and maintenance of his product when combined with other products. Whenever the Manufacturer has specific knowledge of an improper use of his product, he should furnish such information in writing to the Architect. It is recommended that this type of information and communication emanate from the manufacturer's technical staffs, such as application engineering and researching divisions. It is further recommended that the services of public relations and advertising counsel be reserved primarily for improvements in presentation and format of product literature.

When an inquiry is made by an Architect of the Manufacturer concerning an intended use and installation of his product, the Manufacturer is expected to respond with a technically qualified reply.

The Manufacturer is expected to recognize that he is responsible for the failure of his product to perform in accordance with written data supplied by him or his authorized representatives, as well as misrepresentations of such data.

When a Product has been installed in accordance with the Manufacturer's written instructions and written recommendations, and such Product fails, then the Manufacturer has the responsibility therefor. Such responsibility extends to related products affected by the failure, where the Manufacturer has notice of the proposed use of such related products. In case of failure, manufacturers of the other products involved should make available their technical knowledge to the Architect in the correction of the failure.

The Manufacturer is expected to investigate the relation of his product to other components likely or logically expected to be used in association with his. Such information should be available to the Architect.

Architect

The Architect is responsible for proper design. He is expected to inform himself with respect to the properties of the products he specifies, though he is entitled to rely on Manufacturers' written representations. He is advised to seek the technical opinion of the Research or Application Engineering Departments of the Manufacturer when his intended use is not clearly included in the printed data of the Manufacturer. He is further responsible for uses contrary to supplementary written information on proper use and installation procedures of the Manufacturer.

The Architect's use of a Product and its installation should extend to its compatibility with and relationship to adjacent materials and assemblies, notwithstanding the Manufacturer's similar obligations.

Contractor

It is the responsibility of the Contractor to inform himself concerning the application of the products he uses and to follow the directions of the Architect and Manufacturer.

In the event of disagreement between the Contract documents and the Manufacturer's directions, the Contractor is expected to seek written instructions from the Architect before proceeding with the installation.

If the Contractor has knowledge of or reason to believe the likelihood of failure, he is expected to transmit such knowledge to the Architect and ask for written instructions before proceeding with the work.

Owner

It is assumed that the Owner or other person responsible for operation and maintenance of the project will properly maintain the material and equipment in accordance with Manufacturer's recommendations.

Questions for Cogitation

The purpose of this section is to provide food for thought for persons using the book independently and to facilitate classroom discussion when used as a text-book.

Chapter 1

BUILDING INDUSTRY RELATIONSHIPS

1. There is fairly general agreement that the design professions and the constructors constitute two distinct functional segments of the building industry. An alternative classification of industry elements, however, considers the manufacturers of building materials and products to be part of the constructors group rather than a distinct industry component as designated in this book. Identify some factors tending to support this alternative point of view and then state your own preference, giving reasons for your position.

2. Differentiate between private construction and public construction with respect to initiation of construction activity, methods of financing, and building types likely to be involved. How would you classify buildings at private colleges and universities that are partially financed by a government grant?

3. Restate the distinction between investment building and consumption building. How clear-cut is this distinction?

4. Identify factors leading to the conclusion that the bulk of human activity that is carried on indoors must depend on existing facilities. What is the significance of this situation with regard to building activity and to building industry organization?

5. With respect to the characterization of building production as manufacturing, identify some factors supporting this view and some that differentiate the production of buildings from industrial manufacturing generally.

6. Discuss the circumstances that have led to building being characterized as a service industry. What factors within the industry operate to weaken this point of view? Do you think the characterization is a valid one?

7. Restate the five phases of "basic architectural services" as defined by the American Institute of Architects and briefly summarize the activity encompassed in each phase. How do these differ from "project analysis services"?

8. Differentiate between construction by force account and by contract and describe the general contracting method of building construction.

9. Identify the principal virtues of the single contract system and also point out factors that have tended to undermine its dominant position in the building industry.

10. Identify the principal virtues of the separate contracts system and also point out any weaknesses you may think the system possesses.

11. Describe the unit design-construction type of contract frequently known as the "package deal." What are some of the merits claimed for it and some of the disabilities cited against it? Where does a unit design-construction company fit in the classification of industry elements?

12. Differentiate between lump-sum and cost-plus-fee construction contracts, giving the characteristics and principles underlying each type.

13. In connection with building industry ethics and standards, identify some factors that you think may have led to the adoption of

(a) articles 1.3 and 1.5 of the AIA Standards of Professional Practice. (See Appendix A.)

(b) the five items under Article 3 of the "rules" in the AGC Code of Ethical Conduct. (See Appendix B.)

14. Review your understanding of the several stages and events involved in the chronology of a building project from inception to completion.

Chapter 2

CONTRACTS FOR ARCHITECTURAL AND ENGINEERING SERVICES

1. Give your explanation of the concept of single design responsibility as related to building projects. How may it be achieved in the professional practice of architecture?

2. Discuss the merits and/or disabilities of the percentage-of-construction-cost type of A & E contract from the points of view of (a) the owner and (b) the architect-engineer.

3. Discuss the merits and/or disabilities of a lump-sum type of professional services contract from the viewpoint of the A & E firm.

4. Under what conditions would one of the cost-reimbursement types of A & E contracts be more appropriate for a project than either the percentage-of-construction-cost or lump-sum types?

5. Identify the items of professional services that are accounted for by

(a) the multiplier in contracts where compensation is based on a multiple of direct personnel expense; and

(b) the fixed fee in the professional-fee-plus-expense type.

6. Compare the characteristics of lump-sum professional services contracts with those of contracts based on a professional fee plus expense with guaranteed ceiling.

7. What justification is there for establishing architectural fees as a percentage of construction cost (or estimated construction cost)?

8. Make a list of items that should be covered by specific statements when the scope of A & E services is being established for a specific project.

9. Compare the manner of defining scope of professional services used in AIA Document B131 with that contained in the U.S. Government contract form NavFac 4-4330. (Both are reproduced at the end of Chapter 2.)

10. Discuss the nature of customary A & E services during the construction phase of a project that is, what does the architect-engineer undertake to do when he furnishes full construction phase services?

11. Differentiate between the basic services and "additional services" as provided for in A & E contracts. How can you justify the use of these two classifications? How are payments for additional services related to Article 5 of AIA Document B131?

12. Referring to the two standard contract forms reproduced at the end of Chapter 2, define and state the significance of the terms construction cost and net construction cost.

13. Compare the provisions covering the architect-engineer's cost estimate responsibility in the two contract forms presented in Chapter 2.

14. Compare the provisions relating to ownership of plans and specifications in private and federal government professional services contracts.

15. Differentiate between provisions for contract termination for default and for convenience of the owner.

16. Compare provisions for settlement of disputes contained in AIA Document B131 with those in the U.S. Government contract form NavFac 4-4330 (both given in Chapter 2).

17. What is your understanding of the "reasonable care and skill" concept as applied to the professional practice of architecture and engineering?

18. What do you think the circumstances were that led to the promulgation of the AIA's Policy Statement on Building Product Development and Uses? Do you think the divisions of responsibility it advocates are equitable to owner, producer, contractor, and architect?

Chapter 3

ARCHITECT-CONSULTANT CONTRACTS

1. Discuss the matter of an architect's professional responsibility for the work of consulting engineers retained by him.

2. What is normally the extent of a consulting engineer's responsibility to an architect who retains him?

3. Differentiate between the agency and joint venture concepts with respect to architect-consultant relationships. How may the distinction be reinforced by the terms of the consulting contract?

4. What is the significance of designating the architect as project coordinator in a contract for consulting services?

5. Identify four principal types of architect-consulting engineer contracts as related to basis of compensation.

6. Give some reasons supporting the statement that it is generally advisable to use the same type of contract for consulting services as that employed in the prime professional services contract between owner and architect.

7. What justification is there for establishing consulting engineering fees as a percentage of the cost of the consultant's part of the project? Do you think that determination of consultant's fee as a percentage of the architect's fee is a more satisfactory arrangement? What are some of the implications of this scheme?

8. With reference to establishing the compensation of consulting engineers, compare the situation when percentage of construction cost is employed as the basis of payment in both owner-architect and consulting contracts with that when both contracts are of the lump-sum type based on a percentage of the estimated construction cost.

9. How may participation of a consulting engineering firm's principals be accounted for in the multiple-of-consultant's-direct-personnel-expense type of agreement?

10. When the circumstances surrounding a project permit a choice, would you as a consulting engineer prefer a multiple-of-direct-personnel-expense or a fee-plus-expense type contract with the architect? Give reasons supporting your position.

11. What is the basis on which the scope of services is established in architect-consultant contracts?

12. Identify some of the activities normally carried on within the scope of basic services by a consulting engineer (a) prior to award of the contract for construction of a project and (b) subsequent to such award.

13. Review your understanding of additional services by comparing Article 4 of AIA Document C131, with Article 1.3 of AIA Document B131 (given in Chapters 3 and 2, respectively).

14. Restate the provision in AIA Document C131 covering consultant's cost estimate responsibility. Compare this with the architect's similar responsibility as specified in AIA Document B131. Do you think the provisions are consistent and equitable?

15. Review your understanding of items to be covered in the provisions of consulting contracts dealing with termination.

16. With reference to appraising the adequacy of architect-consultant contracts, compare provisions in the three standard contract forms at the end of Chapter 3 as they relate to (a) scope of consulting services; (b) additional services; (c) payments to consultants; (d) architect's responsibilities to consultants; and (e) arbitration of disputes.

Chapter 4
CONDITIONS OF THE CONSTRUCTION CONTRACT

1. State the functions of the following elements of a contract for building construction: (a) General Conditions, (b) Supplementary Conditions, and (c) Agreement.

2. Differentiate between the function of the Drawings and that of the Specifications.

3. Differentiate between Addenda and Modifications as elements of the Contract Documents.

4. Give your understanding of the phrase "the Contract Documents are complementary."

5. In view of the fact that the architect is not a party to the construction contract, what is the purpose of specifying his administrative functions in the General Conditions?

6. Restate the purpose of the architect-engineer's periodic visits to the site during the construction of a project. Check your own thinking against Articles 2.2.4 and 2.2.5 of AIA Document A201 (given in Chapter 4).

7. Describe the architect's position with regard to interpretation of the Contract Documents during the construction phase.

8. Identify some of the points to be considered if a provision for suspension of work at the convenience of the owner is to be included in a construction contract.

9. Define and state the purpose of cash allowances as provided for in the General Conditions.

10. State the responsibilities of contractor and architect-engineer with respect to the processing of shop drawings and samples.

11. Review your understanding of indemnification and hold harmless clauses as treated in Article 4.18 of AIA Document A201. Comment on the purpose and intent of such provisions.

12. Under the provisions of AIA Document A201, what alternative procedures are available to the low bidder when one of his proposed subcontractors is unacceptable to the owner and architect?

13. In view of the fact that no contractual relationship exists between the owner and a subcontractor, what is the purpose of including provisions covering subcontractual relations in the General Conditions (as in Article 5.3 of AIA Document A201)?

14. With respect to disputes arising under private construction contracts, compare the relative merits of resolving them through arbitration or through litigation.

15. Briefly outline the arbitration process, touching on initiation of proceedings, appointment of arbitrators, hearings, and award.

16. Review your understanding of disputes clauses in federal government contracts by reading Article 6 in General Services Administration Standard Form 23-A given in Chapter 4. Do you think this is an equitable provision?

17. Define the term substantial completion and comment on its significance.

18. Define the schedule of values prepared immediately after contract award and describe how it is used.

19. Outline the process by which progress payments are made to the construction contractor.

20. What is the extent of the architect-engineer's responsibility in issuing a certificate for payment?

21. Describe the process leading to final payment under the construction contract.

22. Discuss the architect's latitude of action in making or approving changes in the work.

23. Outline the steps to be followed when the owner desires to make a change in the work, commenting on how the value of additive changes may be established.

24. Identify some reasons that might impel an owner to initiate contract termination proceedings.

25. What are the principal grounds for suspension of work or termination of contract on the contractor's initiative? What are some of the hazards in so doing?

26. Identify the principal additional provisions required in the General Conditions when a project is to be constructed under the separate contracts system.

27. Restate the reasons for dividing the Conditions of the Contract into General and Supplementary Conditions.

28. Identify some of the articles of the General Conditions that normally need additional treatment in the Supplementary Conditions. What additional articles are sometimes required?

Chapter 5

THE SPECIFICATION AS A CONTRACT DOCUMENT

1. Restate the function of the building specifications and identify several points to be considered with respect to the complementary functions of specifications and working drawings.

2. Give your understanding of the division-section concept for organizing building specifications.

3. With reference to the Uniform System, what is the basis of allocating subject matter between Division 1 - General Requirements and the Supplementary Conditions?

4. Identify several subjects treated in the General Conditions that are frequently amplified in Division 1 of the project Specification.

5. With reference to types of technical specifications, compare the principal characteristics of performance specifications and the materials and workmanship type.

6. Differentiate between quality-control testing and performance testing as related to specification requirements.

7. Discuss the matter of technological responsibility with regard to selecting and specifying building materials and components.

8. Identify the key considerations to be taken into account with respect to the use of reference standards in writing building specifications.

9. Differentiate between the open and closed classifications for building specifications.

10. Restate the "or equal" problem as it is encountered in specification writing.

11. With reference to handling substitutions for specified products proposed by bidders and/or the building contractor, describe some alternative methods that might be adopted.

12. Consistency of terms used to denote similar items in the specification sections and on the working drawings is often difficult to achieve. Explain the significance of this problem and indicate some measures that may be taken in an effort to resolve it.

13. With reference to the use of statements of scope in building specifications, cite arguments for and against including articles relating to (a) work included and (b) work not included.

14. Discuss the pros and cons of prefacing the technical specification sections with a statement similar to the following:

> The requirements of the General Conditions, the Supplementary Conditions, and Division 1 of these Specifications apply to all work under this section.

Chapter 6

BONDS AND CONSTRUCTION INSURANCE

1. What is the function of surety bonds as related to contracts for building construction?

2. Review your understanding of the terms Surety, Obligee, Principal, Penal Sum as they apply to a performance bond.

3. Differentiate between performance and payment bonds.

4. Describe the dual-bond system and discuss its merits.

5. Discuss the relation between surety bonds and change orders to the construction contract.

6. Differentiate between a bond and a warranty.

7. Identify five areas of liability normally covered by contractor's liability insurance.

8. What is the purpose of property damage liability insurance as carried by the contractor?

9. Discuss the relationship of contractual liability insurance to contractor's liability insurance generally.

10. State the purpose of owner's liability insurance and indicate its relationship to the contractor's liability insurance.

11. Discuss the coverage contemplated by project and property insurance, commenting on who purchases it and whom it protects as well as the types of hazards normally covered.

12. Describe the procedure for handling payments to the several parties concerned when loss is incurred under project and property insurance.

13. Describe the Builder's Risk Completed Value Form of project and property insurance.

14. Define and state the significance of subrogation as it relates to construction insurance.

15. Describe procedures for assuring that the various parties concerned know that the intended insurance coverage does in fact exist.

Chapter 7

BIDDING AND AWARD PROCEDURES

1. Restate the general nature of the competitive bidding process as related to building construction.

2. Identify the three key bidding documents.

3. Discuss the relative merits of bid bonds and certified checks as bid security. Just what does a bid bond guarantee?

4. Differentiate between an invitation to bid and an advertisement for bids, and identify the principal items of information each should contain.

5. Draft a statement concerning document deposits for inclusion in an invitation to bid.

6. State the purpose of the instructions to bidders and give the nature of information contained therein. What is this document's relationship to the invitation or advertisement?

7. With reference to proposal or bid forms, indicate the type of information called for by the following article titles: (a) bid items, (b) pledges, (c) acknowledgments, and (d) bid security.

8. Define the process of canvassing the bids and identify possible irregularities that are looked for.

9. Comment on the situations when a bidder discovers a mistake in his bid (a) after submission but before the opening and (b) after the opening.

10. How might an owner misuse his reserved right to reject all bids?

11. Distinguish between a notice of award and a notice to proceed.

12. Identify the factors that give rise to the desire for prequalification of building contractors.

13. Give your own thinking on the relative significance of prequalification as it relates to both private and public construction.

14. Identify some of the key items of informatin that are considered pertinent when appraising a contractor's qualifications for a particular project.

15. Cite some arguments used in opposition to the employment of prequalification procedures.

16. Describe the purpose and operation of bid depositories.

17. Differentiate between separate bidding and separate contracts, commenting on the intended purpose of each.

18. Describe the mechanics of the separate bidding system.

19. Identify and comment on some of the factors that are considered to influence the level of bids received on a particular project.

Chapter 8

LUMP-SUM CONSTRUCTION AGREEMENTS AND SUBCONTRACTS

1. Restate the relationship between the Agreement and the other Contract Documents.

2. Restate the types of information usually given in the Agreement under the broad heading of identification of parties and project. Identify the provisions for recording this information in AIA Document A101 (given in Chapter 8).

3. What is the purpose of the detailed enumeration of documents normally cited in the Agreement? How is this information provided for in the standard document referred to in the preceding question?

4. Define contract time and comment on how it is established.

5. State the basis on which liquidated damages provisions may be incorporated in a construction contract. Differentiate between this mechanism for attempting to secure compliance with an established completion date and a "penalty-and-bonus" provision.

6. Define contract sum and identify articles in AIA Document A201 and GSA Standard Form 23-A (both in Chapter 4) that provide for its adjustment.

7. Compare Articles 9.3 and 9.4 of the AIA General Conditions (Document A201) covering progress payments with Article 6 of AIA Document A101 on the same subject. Why is this matter treated in both places?

8. Discuss the purpose and operation of the retained percentage procedure as related to progress payments.

9. How is the retained percentage of progress payments handled in connection with the payment at substantial completion?

10. Review your understanding of the steps leading to substantial completion and final payment as provided for in Article 9.7 of AIA Document A201 (General Conditions). Compare this with Article 7 of AIA Document A101 and state why the subject of final payment is included in the Agreement at all.

11. State the situation with respect to future claims by both owner and contractor after final payment has been effected.

12. Discuss the relation of date of execution of the construction contract to the matter of contract time.

13. Comment on methods for handling identification of contractors in Agreements prepared when a project is to be built under the separate contracts system. Also state how identification of the other pertinent Contract Documents is achieved in the separate Agreements.

14. Comment on provisions relating to contract time and liquidated damages as they may require modification under the separate contracts system.

15. Make a comparison of the short form construction contract represented by AIA Document A107 with the AIA General Conditions (Document A201) and identify specific differences between the two.

16. What is the mechanism usually employed to register the subcontractor's recognition of the provisions of the prime contract?

17. In the prototype subcontract form given in Chapter 8, cite the provisions for obligating the subcontractor to the contract time requirements of the prime contract.

18. State the items to be covered in subcontract provisions relating to progress payments.

19. Identify by article numbers in the AIA General Conditions (Document A201) provisions analogous to the numbered items of Article X of the prototype subcontract form given in Chapter 8.

20. Read the prototype subcontract form in Chapter 8 in the light of Articles 5.3 and 5.4 of the AIA General Conditions (Document A201) and state whether you think the first article is consistent with the requirements of the second article. If you do not think it is consistent, identify the points of apparent inconsistency.

Chapter 9

COST-PLUS-FEE CONSTRUCTION CONTRACTS

1. Discuss the difference in owner-contractor relationships under competitive-bid lump-sum and cost-plus-fixed-fee (CPFF) contracts.

2. Identify some of the advantages and disadvantages of CPFF construction contracts from the owner's point of view.

3. Discuss the situation that arises in connection with identification of contract documents when a CPFF contract is executed ahead of the time complete plans and specifications are available.

4. When establishing the contract time under a CPFF contract, what factors additional to those relating to lump-sum contracts must be taken into account?

5. In view of the underlying principle of CPFF contracts that the contractor shall be reimbursed for all costs incurred in the construction of a project, why are categories of reimbursable and nonreimbursable costs established in the Agreement?

6. In general, just what wages and salaries are allowed as reimbursable costs under CPFF contracts?

7. What items of cost are considered reimbursable with respect to building materials, components, and equipment?

8. Identify some of the items allowed as reimbursable costs under the heading of field facilities.

9. Under CPFF contracts, why does the question of plant and equipment rental from the prime contractor arise?

10. In general, do you think it is in the owner's interest to have the contractor rent the necessary equipment from a third party or for the contractor to furnish the owner by him on a rental basis?

11. In addition to rental charges, what other costs associated with the use of plant and equipment are allowable as reimbursable expenses?

12. How are repair and maintenance of rented plant and equipment handled?

13. Describe the situation relating to "small tool and equipment" costs under CPFF contracts.

14. Compare the provisions relating to payment of bond and insurance premiums under lump-sum and CPFF contracts.

15. Identify some items of a contractor's central organization operating costs that can be allocated to an individual project on a proportionate basis only.

16. Under what conditions can some portion of a contractor's indirect operating costs be allowed as a reimbursable overhead cost?

17. Identify some of the salaries and wages paid by a contractor that are not normally considered as a reimbursable expense under a CPFF contract.

18. On a large project how does establishing a field office on the job modify some of the distinctions between reimbursable and nonreimbursable expenses?

19. What is the problem with respect to allowing overtime pay as a reimbursable expense? What are its implications under a CPFF contract that carries a liquidated damages provision?

20. Even in the most carefully drawn contracts, it is virtually impossible to cover assignment of every cost that may arise to reimbursable or nonreimbursable categories. How can equitable provisions for this situation be made?

21. Identify the two principal components of the fixed fee under a CPFF contract and state some of the considerations taken into account when establishing it.

22. Discuss some of the factors that should be considered when contemplating the incorporation of a guaranteed ceiling cost in a CPFF construction contract.

23. Compare procedures for handling changes in the work under lump-sum contracts and under CPFF contracts both with and without guaranteed ceiling provisions.

24. Give some reasons in support of the view that whenever possible lump-sum subcontracts should be employed under CPFF prime contracts.

25. Describe the process of handling progress payments under CPFF contracts.

26. Differentiate between conventional CPFF construction contracts and the type of cost-plus-fee agreements known as management contracts.

Index

337